KS3
Success

Science

Complete Coursebook

Age 11-14

Nick Dixon
Neil Dixon

Contents

Contents

Contents

Contents

Chapter 18 Electricity and electromagnetism

Chapter 19 Space physics

Scientific attitudes

The principles of science

Scientists try very hard to remain **objective**. This means that they try to not let their personal opinions and beliefs influence the conclusions that they draw from their experiments and data. An important way in which they try to check that they are being objective is to share their results with other scientists and ask them to draw their own conclusions. If the conclusions from the other scientists are the same, this suggests that the conclusions are objective.

Making measurements

It is important to make sure that measurements in science investigations are…

- **Accurate** – this means that they are close to the true value. You should try to avoid any sources of error, like not zeroing the balance before measuring the mass of something.
- **Precise** – this means that multiple measurements of the same thing have similar values. You can check the amount of precision in your investigation by doing repeat readings and looking at the spread of values around the mean.
- **Repeatable** – this refers to whether you get precise (similar) measurements or results if you repeat the procedure several times, using the same equipment and method.
- **Reproducible** – this refers to whether another person would get precise (similar) measurements or results to your results, using a slightly different method and equipment.

The development of theories

Throughout this book you will find examples of how scientific ideas, theories and explanations have changed over time. Change happens when new evidence is discovered or new observations are made that cannot be explained by the older idea or explanation.

A classic example is that, for thousands of years, people believed that the Sun orbited the Earth, because the evidence seemed to suggest that this was true. After all, we have all seen the Sun rise in the sky at sunrise and then fall below the horizon at sunset. In 1543, Copernicus suggested a mathematical model which showed that the Earth and other planets orbited the Sun, and evidence to support this was later found by Kepler (1605) and Galileo (1610). Evidence from Herschel (1790) and other scientists suggested that the Sun was not the centre of the universe. As recently as 1925 Hubble discovered that the Sun is one of several billion stars in one galaxy called the Milky Way, which is just one of over a hundred billion galaxies.

Scientists publish their results, conclusions and ideas for several reasons, including:

- to get credit for them
- so that other scientists can check their results and conclusions
- so that other scientists can build on their work.

Sometimes, scientists put their work on the internet or make presentations at conferences (meetings). However, the best place for a scientist to publish work is in a **journal**. This is a bit like a magazine for scientists. The editors of a journal get each article **peer reviewed**. This means that the article is checked by other experts in the same area of research, to make sure that the conclusions fit the results.

Experimental skills and investigations

Planning and carrying out an investigation

An investigation usually starts with a question or an idea based on an observation of the real world. A **hypothesis** is a statement or proposed explanation for an observation that can be tested using a scientific approach. When a hypothesis has been thoroughly tested, if it can be supported by evidence, it is called a **theory**. Because science never really claims to have a perfect explanation for anything, and there always remains the possibility that something we believe will be proven wrong in the future, we still use the word theory for things that are widely believed by pretty much every scientist to be true. The **theory of evolution by natural selection** is a good example.

Sometimes, data can be collected by **sampling**. This means taking measurements of a population by observing or measuring some of the individuals in it. This is commonly done in biology. For example, you might sample the species present in a stream or field, or you might sample the students in your class to get an idea of how many people in your year at school have different coloured eyes. You need to make sure that you have a sufficiently large sample so that your results are **valid** and reproducible.

Sometimes, an investigation can be a **controlled experiment**. In this type of investigation, you identify an **independent variable** (input), which you change to see what effect it has on a **dependent variable** (output), which you measure. To make this a valid or **fair test**, you need to control as many of the other variables as possible, to prevent them from having an effect on the dependent variable. For example, when investigating how quickly different metal elements react with acid, you would need to control the type of acid used, the temperature, and so on.

Working scientifically

Units of measurement

When doing an investigation, it is important to record your results carefully. Usually, this will involve a results table. If you are taking measurements, the headings for the columns should include units, so you don't need to put units next to the numbers in the table itself. You should use the correct units for a given measurement if possible. The units to use are dictated by the **International System of Units** (SI). Be careful to get your upper and lower case letters correct!

Quantity	SI unit	Other acceptable units	Do not use these units
Distance, length	metre, m	kilometre, km centimetre, cm millimetre, mm light year	inches feet miles
Mass	kilogram, kg	gram, g	ounces pounds stones
Force, weight	newton, N		stone kilogram
Energy	joule, J	kilojoule, kJ	calorie kilocalorie
Current	ampere (amp), A		
Voltage (potential difference)	volt, V		
Speed Velocity	metres per second, m/s		miles per hour
Temperature	kelvin, K	degrees Celsius, °C	degrees Fahrenheit, °F
Time	second, s	minutes	
Pressure	newtons per metre squared, N/m^2 (also called a pascal)	mmHg (for blood pressure only)	PSI
Power	watt, W	kilowatt, kW	horsepower
Resistance	ohm, Ω		

Naming chemicals

It is important for chemists to use the same names to describe chemicals, so that there is no confusion between scientists. Naming elements is easy because they are listed in the periodic table. Writing formulae for compounds is also very clear. But how would you know if sulphurous acid, sulfonic acid and sulfuric(IV) acid are the same thing? The **International Union of Pure and Applied Chemists** (IUPAC) decides on the one single name that any compound should have. Sometimes these names have Roman numerals in them, which you say as a number. For example:

- $CuSO_4$, copper(II) sulfate → You say "copper two sulfate"
- HNO_3, nitric(V) acid → You say "nitric five acid"
- $KMnO_4$, potassium manganate(VII) → You say "potassium manganate seven"

Evaluating risks

It is important to work safely when doing investigations. It is impossible to remove all of the risks but evaluating them before and during an investigation is essential. Here are some of the hazard symbols (pictograms) that you should be aware of when using dangerous substances.

Flammable – keep away from sources of ignition/heat

Corrosive – avoid skin contact, wear gloves

Hazardous to the aquatic environment – do not dispose of down drains

Moderate hazard – for substances that are skin irritants or slightly toxic

Oxidising agent – keep away from fuels

Acutely toxic – do not breathe in, eat or touch

Biohazard – do not allow into contact with living things (e.g. humans)	Health hazards (includes **carcinogens**) – do not breathe in, eat or touch	Radioactive – avoid long term exposure to reduce the risk of cancer

A risk assessment is an important part of planning an investigation. For example, when investigating the thermal decomposition of baking soda:

Hazard (a physical thing that can cause harm)	Risk (what might happen)	Safety precautions (how will you stay safe)	Level of risk (when following precautions)
Hot boiling tube	Could get burned	Use tongs to hold boiling tube	Low
Hot baking soda	It could shoot out of the boiling tube when being heated	Point the boiling tube away from people when heating	Low
Sharp broken glass	Boiling tube could break if dropped or if hot boiling tube is put into cold water	Handle equipment carefully using tongs Do not put hot equipment into cold water Clear away broken glass immediately	Low

Analysis and evaluation

Representing data

A chart or graph is a visual way to represent the results from an investigation. There are three main types that you should know about.

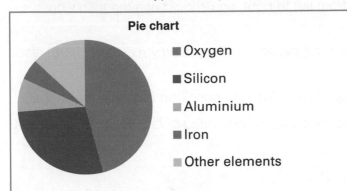

Pie chart
- ■ Oxygen
- ■ Silicon
- ■ Aluminium
- ■ Iron
- ■ Other elements

For representing percentage data

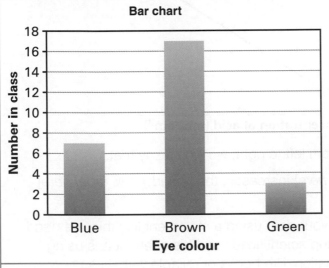

Bar chart

When the independent variable (x axis) is discontinuous data (categories, not numbers)

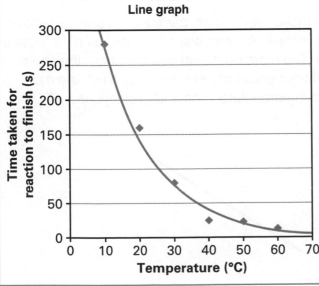

Line graph

When the independent variable (x axis) is continuous data (numerical values that can have any value)

you will find out more about how to use mathematical techniques, such as calculating averages, in the next chapter, **Maths skills for science**.

Drawing conclusions from graphs

Pie charts and bar charts allow you to identify the biggest and smallest values from a set of different categories, but they cannot easily be used to identify patterns and make **predictions**. On the other hand, a line graph can be used for both of these purposes. To draw a conclusion from a line graph that has an upwards trend from left to right, you should use the following general statement:

As the [*insert x axis label here*] increases, the [*insert y axis label here*] also increases.

For example, look at the graph below. We can conclude from this graph that...

As the *concentration of acid* increases, the *speed of reaction* also increases.

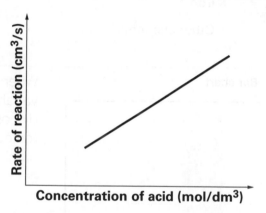

Where the line goes down from left to right, we would say that...

As the [*insert x axis label here*] increases, the [*insert y axis label here*] **decreases**.

After identifying the trend in your data using a statement like this, we need to try to explain the trend using scientific ideas. This often means using secondary sources of information, like books or reliable websites.

We can use this type of line to make predictions, because we can extrapolate the line beyond the range of our data. You can see this below, with the dotted line showing the extrapolation. The more confident you are about the line of best fit (because the points on your graph are all very close to it), the more confident you can be about your extrapolated line and predictions.

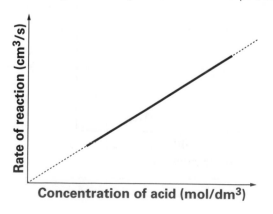

Evaluating your results and method

Looking back at your data, you should try to assess the accuracy, precision, repeatability and reproducibility of your results. Try to think of sources of error in your method. For example:

- Did you zero the balance before measuring any masses (reducing systematic error)?
- Did you read the volumes of liquids in measuring cylinders at eye level rather than from above (reducing systematic error)?
- Did you control all the other variables?
- Were you careful in taking measurements (reducing sources of random error)?

Look at the repeat readings for a given value of the independent variable. Are they similar to each other? This suggests that the results are precise and you can have more confidence in them. Look at how the averages are spread around the line of best fit. The closer they are to the line of best fit, the more confident you can be about your conclusion.

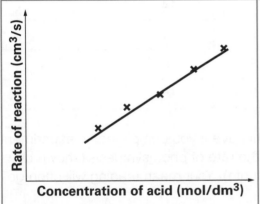 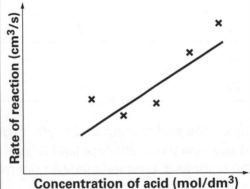

The points are very close to the line of best fit, so you can be confident about the conclusion from this graph.

The points are far from the line of best fit, so you cannot be quite so confident about your conclusion.

Maths skills for science

Averages

The term 'average' has three mathematical meanings:

1. Mean: the average value

2. Median: the middle value when in order

3. Mode: the value that appears most often

Calculating means

It is very unlikely that you will be asked about median or mode in science lessons. However, many science experiments have repeat readings to ensure that the experiments are repeatable. If this is the case it is important we take the mean (or average) reading.

Planning investigations was covered in **Working scientifically**.

Distance from lamp to pondweed (cm)	Number of bubbles released per minute			
	Reading 1	Reading 2	Reading 3	Mean reading
10	13	12	14	
20	8	7	11	
30	6	5	5	
40	4	2	15	

This is the sort of results table you might have if you completed an experiment to see how the intensity of light affects the rate of photosynthesis (shown by the number of bubbles released per minute). Your mean reading will often go in the final column.

To calculate a mean you add the individual numbers together and then divide that total by the number of readings. So here you would do the following calculation:

$13 + 12 + 14 = 39$ and then $\frac{39}{3} = 13$

We divided the total by three because we added three readings together. If we have taken four readings we would have divided the total by four, and so on. Can you calculate the other readings?

Look closely at the reading three for the 40 cm distance. Do you notice anything unusual about it? It is very different from the other two readings at this distance and is likely to be caused by a mistake. Scientists call these **anomalous results**. They are ignored and the experiment repeated. There was not enough time to complete a repeat reading for this distance. How will this affect how you calculate the mean?

It will now be:

$4 + 2 = 6$ and then $\frac{6}{2} = 3$

We only divided by our total by two here because we only had two readings.

Drawing lines of best fit

Line graphs are used to show the results of investigations involving continuous data. These occur when the independent variable (plotted on the x axis) provides data which can have any value. Data that are recorded into groups is called discontinuous (discrete) and would be presented in a bar chart.

In science, we very rarely join up the points in a line graph. We almost always draw a line of best fit. This line does not have to go through all the points, although you will be more confident about your conclusion if it does. If values are a long way from your line of best fit they are likely to be anomalous.

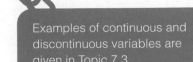

Examples of continuous and discontinuous variables are given in Topic 7.3.

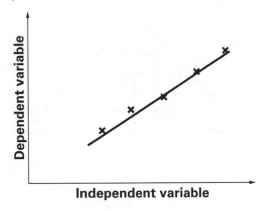

A line of best fit can be a straight line or a curve.

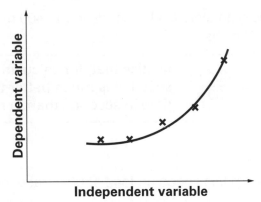

Rearranging equations

We often use equations to calculate readings that are difficult to measure. The equations you need to be able to use are:

- Kinetic energy = ½ × mass × velocity²
- Gravitational energy = weight × height gained
- Energy transferred = power × time
- Work done = force × distance moved in the direction of the force
- Speed = distance / time
- Moment = force × perpendicular distance from pivot
- Pressure = force × area
- Density = mass / volume

Maths skills for science

Questions will often give you the equation you need to use but learning them will mean that you are faster in your tests.

Harder questions often involve rearranging equations. An easy way to do this is to turn an equation into a triangle. Average speed is calculated using the following equation:

$$\text{Average speed (metres per second, m/s)} = \frac{\text{distance travelled (metres, m)}}{\text{time taken (seconds, s)}}$$

The triangle for this equation is:

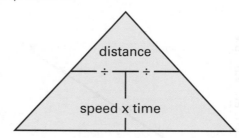

To use a pyramid like this, you put your finger over the thing you want to find out, like speed. Then you can see that you must divide the distance by the time. To calculate the time taken, divide the distance by the speed. To calculate the distance travelled, multiply the average speed by the time taken.

For example, if it takes an athlete 80 seconds to run 400 metres, her average speed is calculated like this:

$$\text{Average speed} = \frac{400 \text{ m}}{80 \text{ s}}$$

(Notice that, for calculations involving speed, it is much better to have the time in seconds than in minutes)

= 5 m/s

If the runner could keep this speed up for a 1500 m race, how long would it take her to finish?

$$\text{Time} = \frac{\text{distance}}{\text{speed}}$$

$$\text{Time} = \frac{1500 \text{ m}}{5 \text{ m/s}}$$

Time = 300 s. You can convert this to minutes by dividing by 60, to give 5 minutes.

Throughout this book, all equations have triangles alongside them to help you use them, except for kinetic energy, which has three terms in it. This means that we cannot put this one into a triangle:

Kinetic energy = ½ × mass × velocity²

Unit conversions

Some values in science have very big numbers. The distance from the Earth to the Sun is 150 000 000 000 metres. Other numbers are very small. The mass of a fly is 0.02 grams.

There are a thousand metres in a kilometre and a thousand grams in a kilogram. This means that if we have a big number we can divide it by 1000 and put 'kilo' before the units. So:

$$\frac{150\,000\,000\,000 \text{ metres}}{1000} = 150\,000\,000 \text{ kilometres (from the Earth to the Sun)}$$

There are a thousand millimetres in a metre and a thousand milligrams in a gram. This means that if we have a small number we can multiply it by 1000 and put 'milli' before the units. So:

0.02 grams × 1000 = 20 milligrams (is the mass of a fly)

Here are some other common examples:

- One thousand joules is a kilojoule, which we often see on food nutritional information.

- One thousand watts make up a kilowatt. The power rating of electrical appliances is often given in kilowatts.

- One thousand milliseconds make up a second. Formula 1 cars sometimes qualify for pole position by milliseconds.

Learning Summary

After completing this chapter you should be able to:
- describe the structure and explain the adaptations of plant and animal cells
- explain how plants and animals are organised
- describe the structure and explain the adaptations of unicellular organisms.

1.1 Plant and animal cells

The **cell** is the fundamental unit of all living organisms. Many plant and animal cells have certain parts, or **components**, in common. Some of these parts are found in both plant and animal cells, whilst others are only in plants.

Generalised plant and animal cells and their components

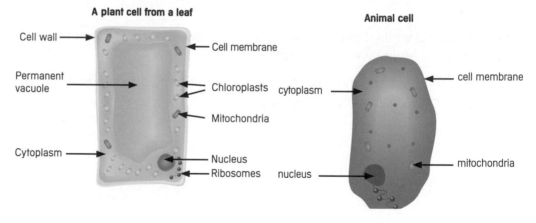

A plant cell from a leaf

Cell wall — Cell membrane
Permanent vacuole — Chloroplasts
Mitochondria
Cytoplasm — Nucleus
Ribosomes

Animal cell

cytoplasm — cell membrane
nucleus — mitochondria

Cell components

Each component of a plant or animal cell has a specific function.

Understanding cell components will help you when you come to respiration (Topic 5.2) and genetic information (Topic 7.1).

Seeing cells

Almost all cells are too small to be seen using your eyes. They are microscopic and so require microscopes to see them. The first microscopes were invented in the 17th century. Scientists could now see inside the cells of larger organisms and could also see single-celled organisms for the first time.

Component	In animals cells	In plant cells	Structure and function
Nucleus	✓	✓	Small circular structure which contains the cell's genetic information (DNA, genes and chromosomes)
Cytoplasm	✓	✓	The liquid inside cells where reactions, including respiration, occur
Membrane	✓	✓	Flexible structure on the outside of cells that controls what enters and exits (the membrane is inside the cell wall of plants)
Mitochondria	✓	✓	Small oval structures found in the cytoplasm where respiration occurs
Wall	✗	✓	Rigid external structure made of cellulose that provides support for the plant cell
Vacuole	✗	✓	Large section in the middle of many plant cells that stores sugars and salts
Chloroplasts	✗	✓	Small green structures that contain chlorophyll for photosynthesis

Make flashcards of the seven types of cell components. On one side draw an image and write whether they are found in plant and/or animal cells. Write down the structure and function of the component on the back of each card.

Plant cells seen using a light microscope

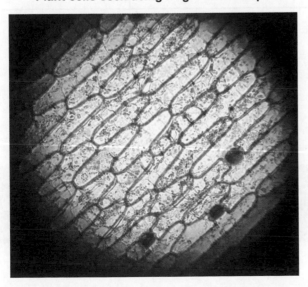

Specialised cells

Many cells in animals and plants have specific functions and are called **specialised cells**. Their structure is adapted for their function.

Understanding specialised cells will help you when you come to look at plants (Topic 4.5) and organisms (Topics 2.1, 2.2 and 2.3)

Cell type	Structure	Adaptation
Sperm cell (animals)		Small, streamlined and contains many mitochondria to swim quickly to find an egg cell Contains **enzymes** in its head to enter an egg cell
Egg cell or ovum (plants and animals)		Large cell that contains an energy store for the young organism when fertilised
Ciliated cell (animals)		Contains tiny hair-like structures (cilia) which wave to remove dirt and bacteria from our gas exchange system
Red blood cell (animals)		Has a biconcave shape and no **nucleus** to increase the surface area to absorb as much oxygen as possible
Palisade cell (plants)		Contains many chloroplasts full of chlorophyll in which photosynthesis occurs
Root hair cell (plants)		Has a long root hair to increase the surface area to absorb as much water as possible

Make a model of a specialised animal cell and a plant cell showing all the different parts of each. You can make your model out of paper or modelling clay, or you could be more creative and use jelly for the cytoplasm and a ping-pong ball for the nucleus!

Progress Check

1. State what components are only present in plant cells.
2. Describe the difference between a cell wall and membrane.
3. Explain how red blood cells are adapted for their function.

1.2 How animal and plant cells make up organisms

Complex organisms like flowering plants or multicellular animals (including humans) are organised in a specific way. The cell is the smallest unit of organisation in all life. Examples are muscle cells in animals and palisade cells in plants.

An animal muscle cell and a plant palisade cell

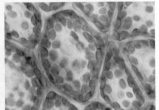

Lots of muscle cells make up muscle tissue and lots of palisade cells make up palisade tissue

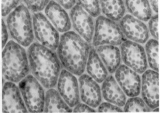

Tissues are lots of identical, specialised cells of the same type in the same place that complete the same function. Examples include muscle tissue in animals and palisade tissue in plants.

Organs are groups of different tissues in the same place that complete the same function. In many animals, muscle and nerve tissues make up the heart organ. Palisade, xylem and phloem tissues in plants make up leaves, which are examples of plant organs.

> You will learn more about muscle in Topic 2.1 and leaves in Topic 5.1.

Muscle and nerve tissues make up heart organs in animals. Palisade, xylem and phloem tissues make up leaf organs in plants

Organ systems are groups of organs that work together to complete the same function. In many animals, the circulatory system is responsible for pumping blood around their bodies. Organisms are made from all their organ systems combined.

> Stick pieces of white paper together to make an area the size of you. Lie down on the paper and ask someone else to draw around you. Research the size and position of your organs. Either draw them onto the outline of your body directly or cut them out of coloured paper and stick them on.

1. State the levels of organisation from cells to organisms.
2. Describe what an organ is.

Progress Check

1.3 Unicellular organisms

Life also exists as simple, **unicellular organisms**. These are one cell in size and so are very small. There are five main groups of unicellular organisms.

Group	Example	Structure
Bacteria	*Streptococcus* which can give you a sore throat	
Archaea	*Methanococcus* which lives near hot volcanic vents on the ocean floor	
Protozoa	*Plasmodium* which is carried by mosquitos and causes malaria	
Algae	*Chlorella* which is a health food	
Fungi	Yeast is used to make bread and beer	

You will learn more about volcanic vents in Topic 6.1.

Understanding yeast will help you when you come to anaerobic respiration (Topic 5.2)

Some unicellular organisms like *Streptococcus* and *Plasmodium* cause diseases. Others are extremely useful to us, like yeast. It is important to remember that not all algae and fungi are unicellular. Seaweeds are multicellular algae and mushrooms are multicellular fungi.

Social and economic implications: Yeast is an economically important organism. We use it to make bread and beer.

The structure of bacteria

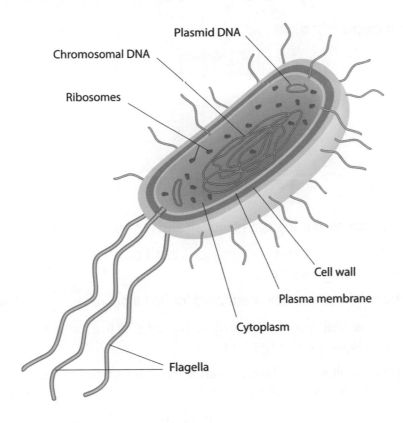

Plasmid DNA

Chromosomal DNA

Ribosomes

Cell wall

Plasma membrane

Cytoplasm

Flagella

Each of the cell components has a specific function.

Component	Structure and function
Chromosomal DNA	The genetic information (deoxyribonucleic acid, or DNA) of the bacterial cells not found in a nucleus like it is in an animal cell
Plasmid DNA	Small closed circles of genetic information (DNA) that can move between bacterial cells
Cytoplasm	The liquid inside a bacterial cell where reactions, including respiration, occur
Wall	Rigid external structure not made of cellulose (unlike plant cells) that provides support
Flagella	Whip-like structures that rotate for movement

Make flashcards of the five types of cell components found in bacteria. On one side draw an image. Write down the structure and function of the component on the back of each card. Add these to the flashcards you made earlier in the chapter.

1. State what a unicellular organism is and give a named example.
2. State what cell components some bacteria possess that other animal cells do not.

Progress Check

Worked questions

a) This is a diagram of a cell.

State two reasons why this is a plant cell and not from an animal. *(2 marks)*

The diagram shows that it has a cell wall and chloroplasts. Only plant cells have these.

b) Describe how a sperm cell is adapted for its function. *(2 marks)*

It has a long tail to help it swim to the egg. It has enzymes in its head to help it enter the egg.

c) Describe two differences between an animal cell from a complex organism and a unicellular organism. *(2 marks)*

Unicellular organisms do not have a nucleus. They also have flagella to help them swim.

d) Explain how complex organisms are arranged. *(4 marks)*

Organisms are made from organ systems. An example is the circulatory system. Organ systems are made from organs. An example is the heart. Organs are made from tissues. An example is muscle. Tissues are made from cells. An example is a muscle cell.

---

a) The answer is awarded one mark for stating that plant cells have a wall and that animal cells do not. The answer is awarded a second mark for stating that only plant cells have chloroplasts. Also stating that only plant cells have a vacuole would have earned a mark.

b) The answer is awarded one mark for describing that it has a tail to help it swim to the egg. It would have been better to use the word ovum here. The answer is awarded a second mark for describing that sperm have enzymes in their heads to help the sperm to enter the ovum.

c) The answer is awarded one mark for describing that unicellular organisms do not have a nucleus. The answer is awarded a second mark for unicellular organisms have tail-like structures called flagella. The term 'tail' is not scientific enough here. A second mark would also have been awarded for describing that unicellular organisms also have plasmid DNA.

d) A mark is awarded for explaining that organisms are made from organ systems and giving an example. A second mark is awarded for explaining that organ systems are made from organs and giving an example. A third mark is awarded for stating that organs are made from tissues and giving an example. A final mark is awarded for explaining tissues are made from cells and giving an example.

Practice questions

1. This question is about specialised plant and animal cells.

    a)

    State the name of this cell. *(1 mark)*

    b) Describe the adaptation of this cell to its function. *(2 marks)*

    c)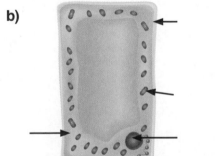

    State the name of this cell. *(1 mark)*

    d) Describe the adaptations of this cell to its function. *(2 marks)*

    e) Explain why a sperm cell might have more mitochondria than an ovum cell. *(3 marks)*

2. This question is about cell components.

    a) State the name of the invention that first allowed scientists to see inside cells. *(1 mark)*

    b)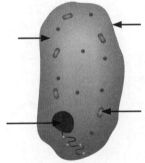

    State the names of the cell components shown by the arrows. *(4 marks)*

    c) Complete the table below to show which cell components are present in animal and plant cells. Some of the answers have been completed for you.

Component	In animal cell	In plant cell
Nucleus	Yes	Yes
Cytoplasm	Yes	
Membrane		
Mitochondria	Yes	
Wall	No	Yes
Vacuole		
Chloroplasts	No	Yes

(6 marks)

    d) Describe the function of a membrane. *(1 mark)*

    e) Explain why plant cells have walls and animal cells do not. *(2 marks)*

3. This question is about arranging organisms.

    a) State the fundamental unit of all life. *(1 mark)*

    b)

This organ is responsible for pumping blood around most animals. State its name and the two types of tissue that make it. *(3 marks)*

    c) Describe how organs make up organ systems. *(2 marks)*

    d) State the name of an organism that does not have organs. *(1 mark)*

4. This question is about unicellular organisms.

    a)

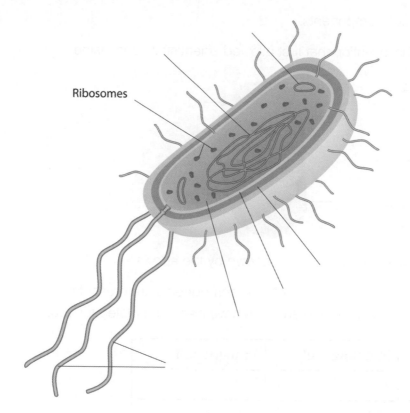

Ribosomes

State the names of the cell components. The first one has been completed for you. *(6 marks)*

    b) Describe the function of the cytoplasm. *(1 mark)*

    c) Explain how unicellular organisms are both harmful and useful. Give the names of two unicellular organisms in your answer. *(2 marks)*

After completing this chapter you should be able to:

• describe the structure and explain the adaptations of the skeletal, muscular, digestive and gas exchange systems in humans.

2.1 The skeletal and muscular systems

The skeletal system

There are 206 **bones** in your body that make up your **skeletal system**. You were probably born with over 300 but some have fused together as you have grown. Bones are organs. The smallest are the three tiny bones that make up your inner ear (the hammer, anvil and stirrup). The longest is your femur (thigh bone) which can be over half a metre long.

The smallest (ear) and largest (femur) bones in your body (not to scale)

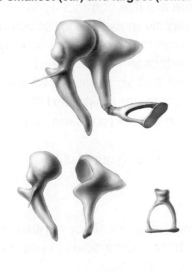

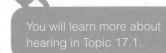

You will learn more about hearing in Topic 17.1.

Functions of your skeleton

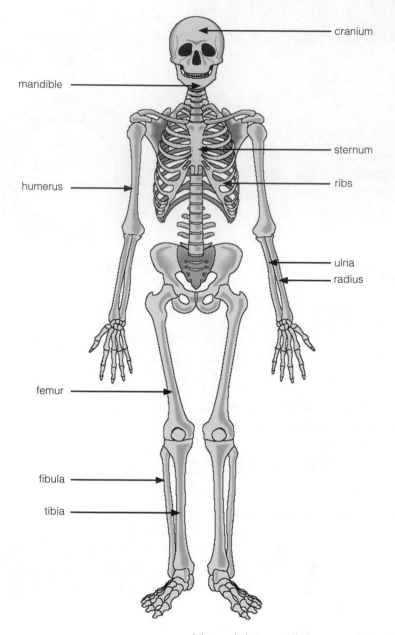

cranium
mandible
sternum
humerus
ribs
ulna
radius
femur
fibula
tibia

Your skeleton:
- supports your **muscles** and internal organs
- protects your internal organs from damage (your skull protects your brain)
- moves when connected to your muscles
- produces red and white blood cells (in the middle of your larger bones).

The bone marrow inside your femur and larger bones makes blood cells

Joints

The places where bones meet are called **joints**. These allow your skeleton to move. There are five common types of joint:
- slightly moveable joints where only small movements occur (between the vertebrae in your spine)
- ball and socket joints where rotation occurs (your hips and shoulders)
- hinge joints, which move like hinges on a door (your knees and elbows)
- pivot joints (your neck)
- fixed joints where movement does not occur (your skull).

The muscular system

Divide a large piece of paper into four equal sized boxes. In each box draw a diagram to represent one of the functions of the skeleton; for example, for movement you could draw a picture of a person running to represent movement.

Muscles are tissues that are found in most animals. They contain protein filaments that can contract. There are about 750 muscles in your body that make up your **muscular system**. This is responsible for:
- movement of your whole body or parts of it
- moving your internal organs (like your heart beating and peristalsis of your digestive system).

Muscles are categorised as voluntary or involuntary. Voluntary muscles such as your triceps and biceps in your arm contract when you want them to. Involuntary muscles such as your heart muscle contract automatically.

Bones and muscles

Bones are connected to muscles by **tendons**. This means that when your muscles **contract** and relax they apply forces to pull your bones into specific positions, which act like levers. This is how you move. The study of how bones and muscles work together is called **biomechanics**. Different muscles in your body apply different forces. It is difficult to make a fair comparison but some of your stronger muscles are in your jaw, your quadriceps in your thigh and your heart muscle. You could measure how strong some of the muscles are in your body using a newtonmeter.

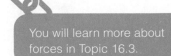
You will learn more about forces in Topic 16.3.

Some of the strongest muscles in a human body

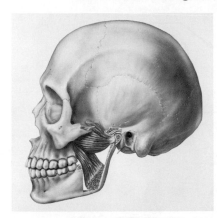

Antagonistic pairs of muscles

Muscles can only relax after they have contracted. They cannot push back with force. So muscles work together in antagonistic pairs opposite each other. When one contracts the second relaxes. When the second contracts the first relaxes. Antagonistic pairs occur in your arms (as shown in the diagram below) and your legs, where your quadriceps and hamstrings work together.

Use drinking straws, a split pin and thread to make a model of bones and muscles in a human arm. Can you identify which bones the straws represent? Instructions are readily available from the internet.

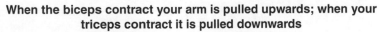
When the biceps contract your arm is pulled upwards; when your triceps contract it is pulled downwards

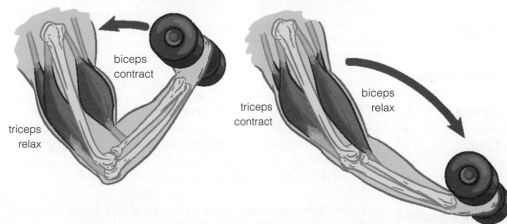

biceps contract

triceps relax

triceps contract

biceps relax

1. State where blood cells are made.
2. Describe how a ball and socket joint works and give an example.
3. Explain how bones act like levers.

Progress Check

2.2 The digestive system

You eat large insoluble molecules of food. These are broken down before being absorbed into your blood and then used. This is called **digestion** and is the function of your **digestive system**.

The eight parts of the digestive system

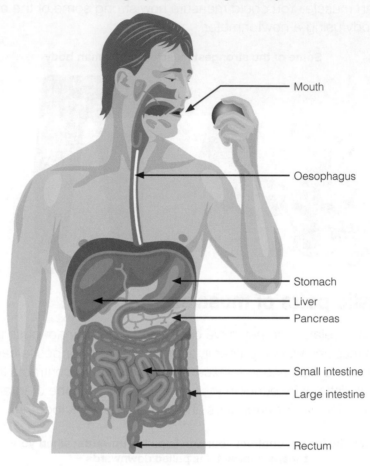

- Mouth
- Oesophagus
- Stomach
- Liver
- Pancreas
- Small intestine
- Large intestine
- Rectum

Put on an old white tee-shirt and find a permanent marker. Stand in front of the mirror and try to draw on where the organs of the digestive system are found. Remember: your reflection in the mirror is reversed. Check your drawing against the one on this page.

Mouth	Teeth mechanically break down food and mix it with saliva for lubrication and add carbohydrase enzymes (amylase) for digestion
Oesophagus	Also known as the food tube or gullet, this organ connects your mouth and stomach
Stomach	Food mixes with acid to kill microorganisms and enzymes for digestion
Liver	Makes bile which breaks down fats we have eaten
Pancreas	Makes the three types of digestive enzyme
Small intestine	Nutrients from food are absorbed into the blood in this organ (even though this organ is called 'small' it is five metres long!)
Large intestine	Water is absorbed into the blood in this short but wide organ, leaving only waste material
Rectum	This is the final section of the large intestine which acts as a temporary store of faeces (undigested food which is passed from your anus)

The total distance that food travels in your digestive system is around nine metres.

Villi in the small intestine

The lining of the small intestine is covered with millions of tiny finger-like projections called **villi**. These structures are about one millimetre tall and increase the surface area of the small intestine in contact with broken-down food to increase its absorption into the blood. This absorption is by **diffusion** and occurs because broken-down food is at a higher concentration in your intestine than in your blood and so moves into it. To do this, villi have a very large blood supply to take away the absorbed nutrients and have a wall only one cell thick to make diffusion easy.

The lining of the small intestine is covered with millions of villi which increase the surface area by about fifty times

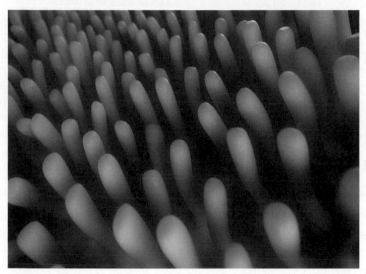

Peristalsis

The lining of much of our digestive system is covered in rings of muscle around it and lengths of muscle along it. This means that it can contract and relax in a rhythmical way to push food through it. This process is called **peristalsis**. We need fibre in our diet to provide something solid for these muscles to push against.

Peristalsis pushes food along our digestive system

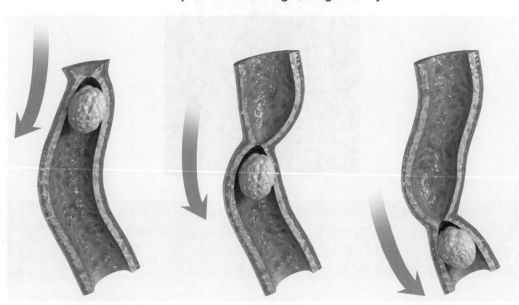

You will learn more about diffusion in Topic 8.1.

Blow up long and thin balloons to make a model of the villi of the small intestine. You could either draw on the red blood vessels or stick on red string (like from the bags you sometimes buy oranges in).

You will learn more about dietary fibre in Topic 3.1.

Push a tennis ball through a pair of tights by squeezing behind the ball each time in a continuous rhythm. This is a model for peristalsis.

Enzymes in the digestive system

Enzymes are often called 'biological catalysts'. They can either speed up the breakdown of large molecules into smaller ones (like those in the digestive system) or help to stick small molecules together. There are three types of enzyme in the digestive system:

Use the corner of your exercise book to make a flicker book showing how an enzyme might break down a substrate.

Type	Location	Substrate it acts on	Breaks substrate into
Carbohydrase	Mouth, pancreas and small intestine	Carbohydrate	Sugars
Lipase	Pancreas and small intestine	Lipids (fats and oils)	Fatty acids and glycerol
Protease	Stomach, pancreas and small intestine	Protein	Amino acids

Bacteria in the digestive system

Make a model of a bacterial cell from your intestine. If you made a model of villi then you could put this model in between them. Don't forget to label all of the bacterial cell components and their functions.

Your stomach has acid in it to kill microorganisms like bacteria before they get into your digestive system and make you ill. However, we have billions of useful bacteria already inside our intestines. We can survive without them but they help us by further breaking down carbohydrates and stopping harmful bacteria from moving in.

There are approximately ten times more useful bacteria in your digestive system than cells in your body

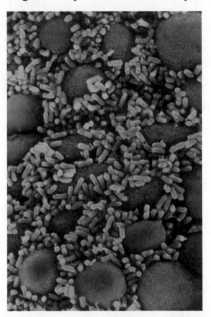

Progress Check

1. State where in your digestive system food first encounters enzymes.
2. Describe the difference in function of your small and large intestine.
3. Explain how peristalsis works.

2.3 The gas exchange system

Our cells require oxygen for respiration. This process produces carbon dioxide and water, which need to be removed. Our **gas exchange system** puts oxygen into our blood and removes carbon dioxide and water from it.

The parts of the gas exchange system

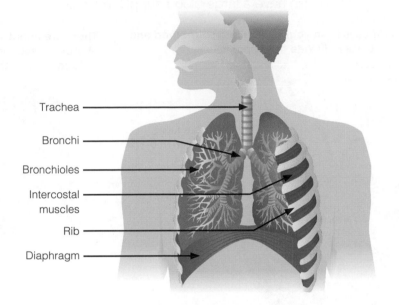

Trachea
Bronchi
Bronchioles
Intercostal muscles
Rib
Diaphragm

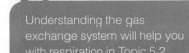

Understanding the gas exchange system will help you with respiration in Topic 5.2.

Make a model of the lungs using an old soft drinks bottle, a sheet of flexible plastic, two straws and two small balloons. What happens when you pull on the sheet at the bottom? What parts of the gas exchange system do the parts represent?

Trachea	This rigid tube leads from your mouth towards your lungs
Bronchi	Your trachea splits into these two tubes which lead into your lungs
Bronchioles	Tiny passages that divide like branches deeper into your lungs
Alveolus	Air sac at the end of bronchiole where oxygen enters and carbon dioxide leaves your blood
Diaphragm	Curved sheet of muscle below your lungs which contracts and relaxes when you breathe
Rib	Protective bones that surround your lungs, heart and other vital organs
Intercostal muscles	Muscles than run between your ribs that contract and relax when you breathe

Rings of cartilage

Your trachea has rings of tough **cartilage** surrounding it. Cartilage is not bone but a flexible and strong tissue (that also makes up your nose and ears). The rings of cartilage keep your airway open even when you are asleep or unconscious. Your oesophagus in your digestive system does not have these and so is not always held open.

If you gently move your fingers up and down the outside of your throat at the bottom of your neck you can feel the rings of cartilage around your trachea.

Alveoli

There are about 500 million alveoli in a pair of adult lungs. These tiny spherical air sacs have a huge surface area about the size of a tennis court. This maximises the transfer of oxygen from your alveoli into your blood and carbon dioxide the other way. This transfer is by diffusion and occurs because oxygen is in a higher concentration in your alveoli than in your blood and so moves from the alveoli into the blood. The lining of your alveoli is moist, which speeds up diffusion. Your alveoli also have a large blood supply to maximum this transfer.

Oxygen diffuses from your alveoli into your blood and carbon dioxide the opposite way

from heart blood low in oxygen

to and from mouth

to heart blood high in oxygen

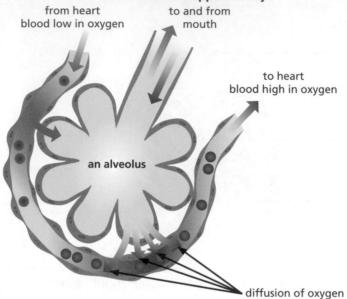

an alveolus

diffusion of oxygen

There are about 170 alveoli for every section of your lungs that is this size

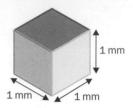

1 mm

1 mm   1 mm

Write each of the stages of breathing in and out on separate cards (for example, ribs move downwards and inwards on one card, volume of lungs increases on another, etc). Mix the cards up and then arrange them into two piles: one for what occurs when breathing in (inspiration) and one for what occurs when breathing out (expiration).

Breathing

Breathing is also called **ventilation**. This process fills our lungs with air and empties them again. It is not to be confused with the chemical reaction of respiration, which occurs in every one of our cells.

When you breathe in:

- Your intercostal muscles move your ribs up and out.

- Your diaphragm contracts and moves downwards.

- Your chest cavity becomes bigger and so has a lower air pressure.

- Air rushes in to fill your lungs.

When you breathe out:

- Your intercostal muscles move your ribs down and in.

- Your diaphragm relaxes and moves upwards.

- Your chest cavity becomes smaller and so has a bigger air pressure.

- Air is forced from your lungs.

Progress Check

1. State the process by which oxygen moves into your blood from your alveoli.
2. Describe the process of breathing in.
3. Explain how your trachea is adapted for its function.

Worked questions

a)

This is a diagram of your femur. State two functions of this bone. *(2 marks)*

This bone makes our red and white blood cells in its marrow. It also helps us to move.

b) Describe how villi are adapted for their function. *(2 marks)*

There are millions of villi in the small intestine. These have a large surface area to absorb food.

c) Describe how your trachea is adapted for its function. *(2 marks)*

The trachea splits into two tubes called bronchi, which enter the lungs. These have rings of cartilage to keep them open.

d) Explain how you breathe in. *(4 marks)*

When we breathe in our ribs go up and out and our diaphragm goes down. This happens because both our intercostal muscles and diaphragm contract. This increases the volume of our lungs and means that air rushes into them.

a) The answer is awarded one mark for stating that the femur makes red and white blood cells and the fact that this happens inside the bone. A second mark is awarded for stating that the femur helps us to move. An alternative mark could have been awarded for stating that the femur supports us.

b) The answer is awarded one mark for stating that the villi are present in the small intestine. A second mark is awarded for stating that they have a large surface area. An alternative mark would have been awarded for stating that they have a large blood supply to maximise diffusion of food into the blood.

c) The answer is awarded one mark for describing that the trachea has tough rings of cartilage around it. A second mark is awarded for describing that these keep the airway open at all times.

d) The answer is awarded one mark for explaining that our ribs go up and out and that our diaphragm goes down. A second mark is awarded for explaining that this happens as a result of the intercostal muscles and diaphragm contracting. A third mark is awarded for explaining that this increases the volume of our chests. A final mark is awarded for explaining that air rushes in.

Practice questions

1. This question is about the skeletal system.

   a) 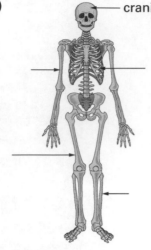 State the names of these bones in this skeleton. The first one has been completed for you. *(4 marks)*

   cranium

   b) Joints occur where bones meet. State their function. *(1 mark)*

   c) Describe how a hinge joint works and give an example in your answer. *(2 marks)*

   d) Add a label to the diagram above to show where fixed joints occur in the human body. *(1 mark)*

   e) Describe how specific parts of the skeleton protect vital organs. *(2 marks)*

2. This question is about the muscular system.

   a) State what action muscles can complete. *(1 mark)*

   b) State the names of the two muscles that allow you to raise your lower arm. *(2 marks)*

   c) Describe how bones and muscles work together to allow movement. *(3 marks)*

   d)

   iris

   This is a photograph of your iris muscle, which controls the size of your pupil. State the name for this type of muscle and describe why muscles act in this way. *(2 marks)*

   e) 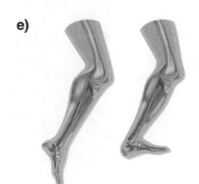 This is a diagram of the major muscles in your leg. State the name for pairs of muscles like this and explain how they work. *(4 marks)*

3. This question is about the digestive system.

a)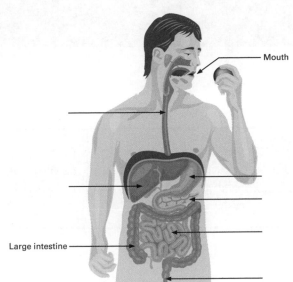
Label the parts of the digestive system. Two have been completed for you. *(6 marks)*

Mouth

Large intestine

b) Describe the two functions of the stomach. *(2 marks)*

c) Describe how carbohydrase enzymes help digestion. *(3 marks)*

d) Describe the action of peristalsis. You can draw a diagram to help you. *(2 marks)*

e) Explain how and why digested food moves into the blood in the small intestine. *(3 marks)*

4. This question is about the gas exchange system.

a)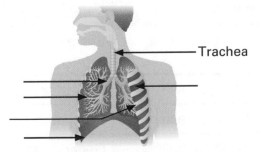
Label the parts of the gas exchange system. One has been completed for you. *(5 marks)*

Trachea

b)
Add arrows on this diagram to show the movement of substances into and out from the blood. *(3 marks)*

c) Explain why this movement occurs. *(2 marks)*

3 Health

Learning Summary

After completing this chapter you should be able to:
- describe a healthy diet and explain the consequences of an unhealthy diet
- describe the effects of exercise, smoking, asthma and drugs on your body.

 7

3.1 A healthy and balanced diet

To live a healthy lifestyle we should:
- eat a balanced diet and not consume more energy than we need
- exercise regularly
- not drink excessive alcohol, smoke or take other drugs.

There are six food groups:

Food group	Found in	Used for or why needed
Carbohydrates	Bread, pasta, rice, potatoes	Energy
Proteins	Meat, eggs, fish	Growth and repair
Lipids (fats and oils)	Butter, oil, cream	Store of energy
Vitamins	Vegetables, fruit	General health
Minerals	Iron in red meat	Healthy blood
	Calcium in milk	Strong teeth and bones
	Sodium in salt	Healthy nerves
Dietary fibre	Vegetables, fruit	Healthy digestive system

This is linked to cell components in Topic 1.1.

Water is often included as a food group but is not really a food. Three-quarters of you is made up of water. All of the chemical reactions needed to keep you alive occur in the cytoplasm of your cells. This is almost all water.

We need food from each of these groups to stay healthy

Balanced diet

Different people need different amounts of these food groups to be healthy. A **balanced diet** provides this. The diagram below gives you a general indication of how much you need each day to be healthy.

Split a paper plate up into the same proportions as the pyramid on the left. Draw or stick on pictures of the foods in the appropriate sections.

Energy requirements

Different foods have different amounts of **energy** in them. We can tell how much by looking at the nutritional information sections of food packaging.

The nutritional information from a food label

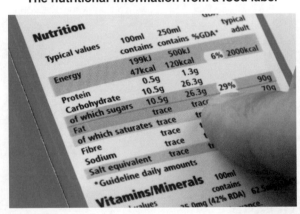

Energy is often measured in calories but the correct units are joules or kilojoules (1000 joules make up a kilojoule). Different people require different amounts of energy each day:

Keep a food diary of everything you eat for the next 24 hours. Is your diet a balanced one? If you are able to, cut out and keep the dietary information sections from your food. Is your energy intake about right?

	Recommended daily intake (calories)	Recommended daily intake (kilojoules)
Man	2500	10500
Woman	2000	8350
Child	1800	7500

These values can vary though, depending upon your age, how active you are and your size.

Unhealthy or imbalanced diets

Your diet can be unhealthy because:
- you eat too much or little food generally, or
- you have the correct amount of food but the wrong amounts of each food group.

Eating too much food and not exercising enough can lead to an increase in weight and eventually to **obesity**. People who are obese are very overweight and have a lot of body fat. They often find it difficult to exercise. Obesity leads to other health problems, including:
- type 2 diabetes (where you can't regulate the sugar in your blood)
- heart disease
- some cancers
- strokes.

Too little food can lead to weight loss, malnutrition and eventually starvation. It can also lead to:
- always feeling tired
- reduced immunity to infection
- poor concentration
- depression.

Obesity and malnutrition can lead to short- and long-term health problems

Make flashcards of the components of a healthy diet, with the name of the component on one side and what it is required for on the other. Use these to revise this topic and then test yourself.

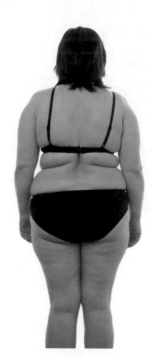
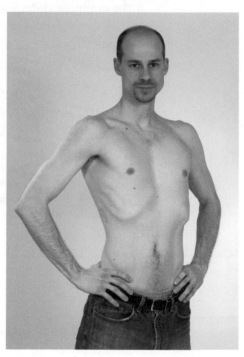

Deficiency diseases

A **deficiency disease** occurs when your body does not have enough specific vitamins or minerals.

Vitamin or mineral	Disease	Symptoms
Calcium	Ricketts	The softening of bones, which can lead to fractures and deformity
Vitamin C	Scurvy	Tiredness, spots on thighs and legs and painful, bleeding gums
Iron	Anaemia	Tiredness, weakness and paler skin

> Make flashcards of the deficiency diseases on these pages. On one side of a card write the name of the disease and the key vitamin or mineral. On the other side of the card write the negative health effects of the disease. Use your flash cards to revise and then test yourself.

Sailors in the 17th century often suffered from scurvy because they did not eat enough fresh fruit on their long voyages.

Social and economic implications: Living healthy lifestyles to reduce the incidences of being ill is important for both social and economic reasons.

1. State what food group you should normally eat most of.
2. Describe what can happen if daily energy requirements are exceeded.
3. Explain why we need to eat dietary fibre even though we cannot digest it.

Progress Check

3.2 A healthy gas exchange system

You will learn more about your gas exchange system in Topic 2.3.

Your gas exchange system puts oxygen into your blood and removes carbon dioxide and water from your blood. It consists of your airways, lungs and surrounding muscles.

Exercise

Regular exercise increases the overall amount of air you can breathe in. It also:

- increases the strength of your heart and other muscles
- reduces your risk of heart disease
- reduces your resting heart rate and quickens your recovery rate after exercise
- improves flexibility and strength of tendons and **ligaments**.

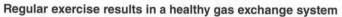

Regular exercise results in a healthy gas exchange system

Smoking

Smoking causes cancer of the lungs, mouth and throat. It breaks down the walls of the alveoli, reducing the surface area and making gas exchange harder. Smoking also damages the cilia cells which line the airways to move mucus out of the lungs, causing coughing. This can eventually lead to **chronic obstructive pulmonary disease** (COPD). It also:

- increases the chances of heart disease, strokes and other cancers
- reduces your fertility
- can prematurely age your skin.

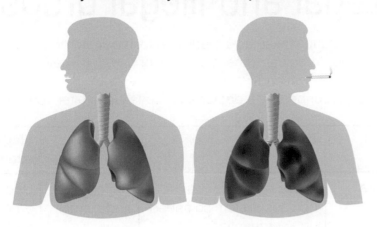

Smoking results in an unhealthy gas exchange system and many other health problems

Social and economic implications: Smoking regularly costs a lot of money and many people these days find it an antisocial habit.

Asthma

Almost one person in ten in the UK suffers from **asthma**. It is a long-term condition that cannot be cured. We are not really sure of its cause and its severity varies from person to person. It results in inflammation of the airways, which makes it harder to breathe. This leads to coughing, wheezing and breathlessness. An asthma attack is a severe onset of these symptoms and can be triggered by pollen, dust, animal fur, cigarette smoke and chest infections, amongst other things. Treatment involves changes to lifestyle as well as medicines.

Inhalers are often used to treat the symptoms of asthma

1. State where smoking is likely to cause cancer.
2. Describe what effects regular exercise has upon heart rate.

Progress Check

3.3 Legal and illegal drugs

A **drug** is a substance taken as a medicine, to intoxicate or to enhance performance. So drugs can either be useful or harmful. They can be grouped according to their effect and whether they are legal or illegal. Recreational drugs are not prescribed by a doctor and are taken to alter your mood.

Group	Effect	Legal examples	Illegal examples
Stimulants	These stimulate your nervous system so speed up your reactions They can make you feel tense or anxious	Caffeine and nicotine	Cocaine, amphetamine and ecstasy
Depressants	These do not make you feel depressed but depress your nervous system so slow your reactions They can make you feel irritable, aggressive and confused	Alcohol	Barbiturates and cannabis
Hallucinogens	These affect your brain so change your perception and result in hallucinations They can make you scared and paranoid	None	LSD
Painkillers	They stop pain signals travelling along nerves to your brain They can have side effects including dizziness or rashes	Paracetamol	Ketamine

Make flashcards of the legal and illegal drugs on these pages. On one side of a card write the name of the drug. On the other side of the card write the negative health effects of the drug. Use your flash cards to revise and then test yourself.

Many medicines can only be taken if prescribed by a doctor. These must only be taken by the person they have been prescribed for. Illegal drugs are grouped according to the severity of their effects on the body:

- Class A: these include heroin, cocaine, ecstasy and LSD.
- Class B: these include cannabis, amphetamine and barbiturates.
- Class C: these include anabolic steroids and ketamine.

The most severe punishments by police and the courts are for class A drugs, then class B and finally C. All are illegal, however.

Cigarettes

Cigarette smoke contains over 4000 different chemicals, of which many are toxic, harmful to your health or cause cancer. Cigarettes contain nicotine, which is an extremely addictive legal stimulant. The tar in cigarettes is one of the major cancer-causing chemicals. It also damages the airways of your gas exchange system. Carbon monoxide is a poisonous gas in cigarette smoke. This takes the place of oxygen in your blood, making you light-headed or breathless.

Smoking cigarettes causes cancer and damage to your gas exchange system

Alcohol

Alcohol is an addictive legal depressant. Excess alcohol in the short term can lead to blurred vision, drunkenness, vomiting and antisocial behaviour. The long-term effects of alcohol abuse are brain and liver damage, which can result in unconsciousness, coma and death.

Social and economic implications: Drinking excessively costs a lot of money and can result in antisocial activities.

Solvents

Solvents are chemicals that dissolve solutes. Some people inhale these harmful chemicals. They can cause instant and permanent damage to your lungs, liver, brain and kidneys. They can also cause hallucinations and change your behaviour or personality.

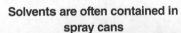

Solvents are often contained in spray cans

Misuse of drugs

Taking too much of a drug such as alcohol or using drugs in ways they are not designed for (such as inhaling solvents) is called drug misuse. This almost always has negative consequences for society or your health. People who are addicted to drugs are called addicts.

You will learn more about solvents in Topic 9.2.

1. State what group of drugs caffeine and ecstasy belong to.
2. Describe the consequences of drug misuse.
3. Explain what effects depressants have upon your nervous system and why their use can be dangerous.

Progress Check

Worked questions

a)

State the three food groups on this plate. *(3 marks)*

Fish is protein. Chips are carbohydrate. Peas have lots of vitamins in them.

b) Describe why we need to eat protein. *(1 marks)*

Protein helps us grow and repair ourselves.

c)

Sailors in the 17th century often suffered from a deficiency disease. State the name of this disease, describe its symptoms and explain why sailors often suffered from it. *(3 marks)*

The disease is called scurvy. If you have this disease you often feel tired and have painful bleeding gums. Sailors often suffered because they did not have enough vitamin C in their diet when they were at sea.

d) Explain what effects smoking has upon your gas exchange system. *(3 marks)*

Smoking causes mouth, throat and lung cancer. It breaks down the walls of the alveoli reducing the surface area. This makes it harder to breathe.

a) The first mark is awarded for stating that fish is protein. The second mark is awarded for stating that chips are carbohydrate. A final mark is awarded for stating that peas provide vitamins.

b) The answer is awarded one mark for stating that protein is used for growth and repair.

c) The answer is awarded one mark for stating that the deficiency disease was scurvy. A second mark is awarded for saying the symptoms are tiredness and painful bleeding gums. This mark could have been awarded for describing spots on thighs and legs. A final mark is awarded for stating sailors often suffered from this disease because they did not eat enough vitamin C.

d) The answer is awarded one mark for mentioning cancers of the gas exchange system. A second mark is awarded for explaining that smoking breaks down the walls of the alveoli. A final mark is awarded for stating that this makes breathing harder. Alternative marks would be awarded for explaining about damage to the ciliated cells giving smokers coughs or mentioning COPD.

Practice questions

1. This question is about leading a healthy lifestyle.

    a) State two benefits of regular exercise.　　　　　　　　　　*(2 marks)*

    b) Describe the symptoms of asthma and how we treat it.　　　*(2 marks)*

    c) Describe what a balanced diet is and explain what can happen if we don't have one.　　　　　　　　　　　　　　　　　*(3 marks)*

    d)

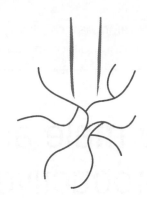

   This is a diagram of an alveolus from the lungs of a non-smoker. Draw a diagram to show what effects smoking has upon an alveolus of a smoker.　*(2 marks)*

    e) Explain the effects of smoking on alveoli.　　　　　　　　*(5 marks)*

2. This question is about drugs.

    a) State what drugs are.　　　　　　　　　　　　　　　　*(1 mark)*

    b) State the name of the group of drugs that includes caffeine and cocaine.　*(1 marks)*

    c) Describe the effects that drugs like caffeine and cocaine have upon the body.　*(2 marks)*

    d) Describe a difference between the legal status of caffeine and cocaine.　*(1 mark)*

    e) Draw lines and label body parts to show where both solvents and alcohol affect the human body.　　　　　　　　　　　　　　　　*(4 marks)*

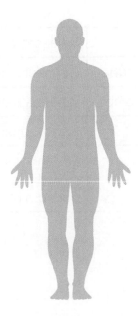

4 Reproduction

Learning Summary

After completing this chapter you should be able to:
- describe the structure and explain the adaptations of the reproductive systems
- describe the development of a baby from fertilisation to birth
- describe sexual reproduction in other animals and reproduction in plants.

 4.1 The male and female reproductive systems

Female

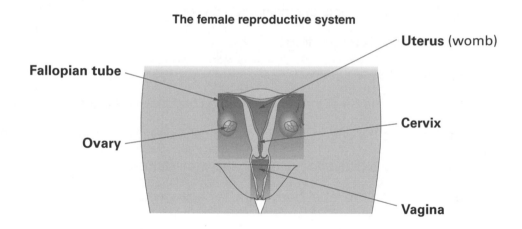

The female reproductive system

Uterus (womb)

Fallopian tube

Ovary

Cervix

Vagina

Ovary	Releases an ovum (egg) every 28 days
Fallopian tube	Connects the ovaries with the uterus
Uterus (womb)	Where a baby develops during pregnancy
Cervix	Narrow opening between the uterus and vagina
Vagina	A man's penis enters here to ejaculate sperm during sexual intercourse and a baby passes through here during birth

Male

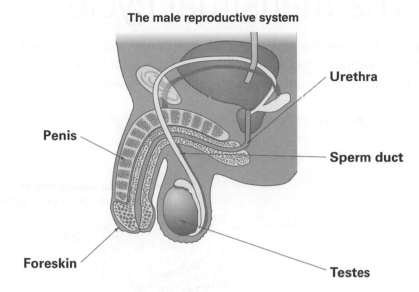

The male reproductive system

Urethra

Penis

Sperm duct

Foreskin

Testes

Testes	Produce sperm
Sperm duct	Connects the testes to the urethra
Urethra	Carries sperm and urine from the body at different times
Penis	Passes urine from the bladder and becomes erect to ejaculate millions of sperm into the vagina during sexual intercourse
Foreskin	Covers the end of the penis when not erect
Scrotum	Holds the testes slightly away from the body to keep them cooler providing the best conditions for sperm production

Draw a simple diagram of the male and female reproductive system without labels on a piece of paper. Cut out smaller pieces of paper to use as labels and write on the names of the parts of both. Then without looking at this page, arrange the labels next to the correct parts of the systems.

1. State the function of the fallopian tube.
2. Explain why testes are held in the scrotum.

Progress Check

4.2 The menstrual cycle

Between the approximate ages of 13 (after **puberty**) and 50, women undergo a regular cycle approximately every 28 days unless they become pregnant. This is called the **menstrual cycle** and prepares the woman's body in case she does become pregnant. This cycle is controlled by the **hormones** oestrogen and progesterone.

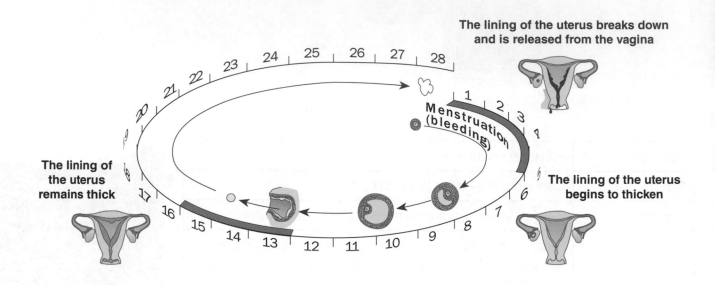

Approximate days	What happens
Days 1 to 4	Bleeding starts as the lining of the uterus is released from the vagina (called menstruation or having a **period**)
Days 4 to 14	The blood vessels lining the uterus thicken in preparation for a fertilised **ovum** (egg) to embed and develop into a baby
Day 14	An ovum is released from an ovary and so women are most likely to become pregnant on or just after this day
Day 14 to 28	The lining of the uterus remains thick in preparation for the fertilised ovum to embed and develop into a baby

Progress Check

1. State how the menstrual cycle is controlled.
2. Describe what happens from days 4 to 14 of the menstrual cycle.

From fertilisation to birth

Fertilisation

Male reproductive cells are called **sperm** and female reproductive cells are called ova (eggs). Together sperm and ova are called **gametes** or sex cells. These special cells contain exactly half of the genetic information of almost all other body cells. When a sperm meets an ovum in the fallopian tube, the two sets of genetic information join together. This process is called **fertilisation** and is the beginning of a new life.

Sperm swimming along a fallopian tube towards an ovum

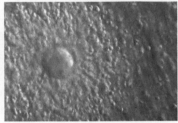

After release from an ovary, this ovum is floating along a fallopian tube

Only one sperm can fertilise an ovum to form a new life

This topic is linked to specialised cells in Topic 1.1 and genetic information in Topic 7.1.

Becoming pregnant

A fertilised ovum is called a **zygote**. About a day after fertilisation this single cell will divide into two and then keep on doubling. After several more days it becomes an **embryo**. After about a week it will have floated to the uterus. It must embed itself into the lining to develop into a **fetus** and then baby.

Make models of sperm and ova cells. Remember, ova are much larger than sperm cells. Label each with its parts and adaptations.

Carrying a baby (gestation)

Time after fertilisation (months)	Size (cm)	Extra information
One	1	The embryo has many major organs including a heart and brain
Two	5	The fetus can move and its eyes have begun to form
Three	15	The fetus has developed fine hair on its head
Five	30	The fetus now has developing fingerprints and can open its eyes
Seven	45	The fetus could just survive if born from now onwards
Nine	50	The baby is now ready to be born

An average baby in the UK has a mass of just over three kilograms before it is born. Try putting three bags of sugar into a rucksack and wearing it low down and backwards (so around your belly) for a day to see how physically hard it is to be pregnant.

A 4-week embryo

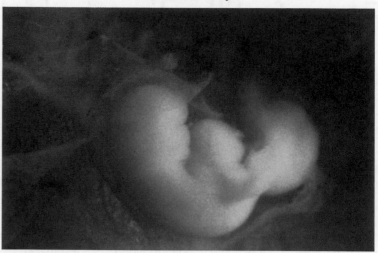

The placenta

After the embryo has embedded itself into the lining of the **placenta**, it is entirely dependent upon its mother. The fetus is protected by the uterus and the **amniotic fluid** within it. This fluid surrounds the baby in a bag called the **amnion**. The placenta is an organ that develops in the uterus only when a woman is pregnant. It is joined to the fetus by the **umbilical cord**. Through this cord, all the oxygen and nutrients are passed from the mother's blood to the baby. The baby and mother do not actually share the same blood but oxygen and nutrients move from the mother's blood to that of the fetus. This movement is by diffusion and occurs when substances are in a higher concentration in the mother's blood than in the babies' and so move into it. Waste products, including carbon dioxide, diffuse back the other way.

If a pregnant woman chooses to lead an unhealthy lifestyle by smoking cigarettes or drinking alcohol for example, the **toxins** she takes in are likely to be directly passed to the baby. This can have adverse effects on the baby, including being born prematurely, at the right time but underdeveloped or dead (a miscarriage).

You will learn more about diffusion in Topic 8.1.

Topic 3.3 covers more about cigarettes and alcohol.

The umbilical cord joins the mother's placenta to the baby's belly button

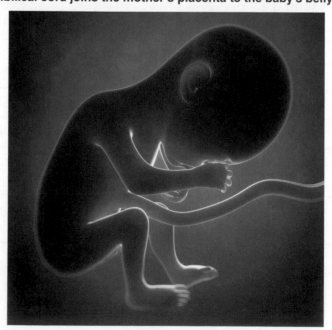

The amniotic fluid surrounds and protects the unborn baby

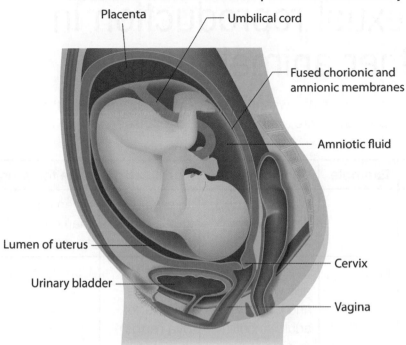

Placenta

Umbilical cord

Fused chorionic and amnionic membranes

Amniotic fluid

Lumen of uterus

Cervix

Urinary bladder

Vagina

An early warning sign of being close to birth is when the amnion breaks, releasing the amniotic fluid out through the vagina. We say that a woman's waters have broken when this occurs. Powerful **contractions** of the muscles of the uterus push the baby through the cervix and vagina. The midwife then cuts the umbilical cord close to the baby's belly button. The baby will now breathe air for the first time and usually start to cry. Finally, the placenta and the rest of the umbilical cord pass out from the mother's vagina and labour has finished.

Immediately after birth the baby is still connected to the mother's placenta by the umbilical cord

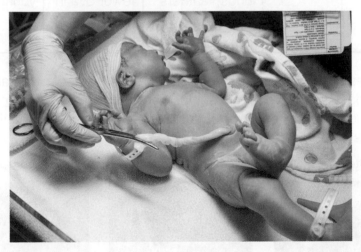

Put a chicken egg into an empty jam jar. Fill the jar completely with water and screw on the lid. Shake as hard as you can to try to break the egg. This is a model for the fetus, amniotic fluid and sac.

1. State the two types of gametes in humans.
2. Describe how a fetus is protected.
3. Explain how a bad lifestyle can affect the development of a fetus.

Progress Check

4.4 Sexual reproduction in other animals

Animals are divided into those with backbones (**vertebrates**) and those without (**invertebrates**). There are five groups, or classes, of vertebrate. Humans are mammals.

Class	Example	Characteristics	Fertilisation	Care for young
Mammal	Cat	Warm-blooded, lungs, hair or fur	Internal, gives birth to live young	Often for longer than other animals
Amphibian	Frog	Cold-blooded, gills when young and lungs when adult, smooth and moist skin	External, lays large number of ova (eggs) in water	Usually none
Reptile	Snake	Cold-blooded, lungs, dry scales	Internal, lays fewer ova on land	Usually none
Fish	Shark	Cold-blooded, gills, wet scales	External, lays large number of ova in water	Usually none
Bird	Goldfinch	Warm-blooded, lungs, feathers and beak	Internal, lays fewer ova in nest or on land	Often until young fly the nest

Make flashcards of the five classes of vertebrate and how they reproduce. On one side of a card write the group of vertebrate and their characteristics. Don't forget to add some examples. On the other side of the card write where fertilisation occurs and how they care for their young.

Invertebrates include insects, worms, crabs, snails, octopi and make up 96% of animal species. Like vertebrates, most reproduction is sexual so involves the production of gametes. Invertebrates possess a range of different reproductive strategies.

Kittens

Newly hatched chicks

Frogspawn

Progress Check

1. State what other animals besides mammals reproduce internally.
2. Explain why amphibians and fish lay larger numbers of ova.

4.5 Reproduction in plants

Plants can either reproduce sexually by the formation of gametes followed by fertilisation or they can make genetically identical copies of themselves. We call this **asexual reproduction**.

Asexual reproduction

Spider plants produce tiny plantlets on runners, which eventually snap leaving a new plant a short distance from the parent. There is only one parent plant and the offspring are genetically identical to each other and their parent.

Sexual reproduction

Flowers are the reproductive organs of many plants.

Stamens	The male part of the flower. Made of anthers containing **pollen** held on a **filament**. Pollen contains the male gamete.
Carpels	The female part of the flower. Made of the **stigma**, **style** and **ovary**. Ova in the ovaries are the female gamete.
Petals	Often brightly coloured to attract insects for pollination
Sepals	Often green and found below the flower. They protect young flowers when they are just buds.

Collect as many different flowers as you can. Carefully place them between several sheets of kitchen roll in the middle pages of a thick, heavy book. Place other books on top and leave them for several weeks. You will then have pressed dried, preserved flowers.

Pollination

Pollination occurs when the male gamete, pollen, meets the female gamete, ova (eggs). Pollen from the **anther** must reach the carpel. It first touches the stigma. Here, a pollen tube is formed, which grows down through the style to the ovary. The nucleus of the pollen cell then moves down the tube to the ovary and fertilises the ovum. This forms a seed. The ovary then develops into a fruit and surrounds the seed.

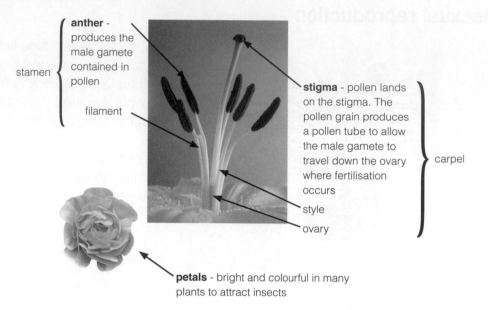

anther - produces the male gamete contained in pollen

stamen

filament

stigma - pollen lands on the stigma. The pollen grain produces a pollen tube to allow the male gamete to travel down the ovary where fertilisation occurs

carpel

style

ovary

petals - bright and colourful in many plants to attract insects

Self-pollination occurs when pollen from the anther is transferred to the stigma of the same plant. Cross-pollination occurs when pollen from the anther of a different plant is transferred to the stigma. Insects often transfer pollen between plants. They are attracted to flowers with bright petals or strongly scented **nectar**. Insects drink the nectar as a source of food. Pollination can also be by the wind blowing it from the anther to the stigma of the same or different flowers.

Social and economic implications: We rely heavily upon insects to pollinate our crops and would not have a secure amount of food to eat without them.

> Take photos of different flowers you see in your garden/neighbourhood. Divide the photos into those that seem adapted for wind pollination and those adapted for insect pollination.

Bees often transfer pollen from one flower to another as they feed on the nectar

Plants which are pollinated by the wind often have long filaments to allow the pollen to be blown away and large feathery stigmas to catch the pollen

Seeds and dispersal

Seeds are formed when the nucleus of a male pollen cell fertilises a female ovum. All seeds have an embryo inside them, which will start to grow into a new plant (**germinate**) when the conditions are correct. They also have a food store to provide the initial energy needed for germination and growth. They are surrounded by a protective coat.

Seeds can be spread (**dispersed**) by the wind as with sycamore trees. They can also be dispersed by water, as with water lilies and coconuts. They can be dispersed by animals, usually birds and mammals. Some fruits are eaten by animals, which release the undigested seeds in their droppings. Other seeds have hooks that attach to passing animals and are dispersed when they fall off. Finally, seeds can be forcibly ejected from pods when they dry out, as in many pea plants.

Seed dispersal is important so that the newly formed plants don't grow close to their parents and compete for resources. Some plants produce many smaller seeds for dispersal like dandelions. Others, like coconut palms, produce fewer, larger seeds.

Dandelion seeds are dispersed by the wind

Coconuts are dispersed by water

Take photos of different flowers you see in your garden/neighbourhood. Divide the photos into those that are adapted to disperse their seeds in the five methods given on this page.

Birds eat blackberry fruits and disperse their seeds in their droppings

Burdock seeds have hooks on which attach to passing animals

Peas eject their seeds from their pods

1. State what makes plant reproduction different from the way most animals reproduce.
2. Describe what happens when pollen reaches the carpel.
3. Explain the strategies that flowers use to attract insects and why they have these.

Progress Check

Worked questions

a) A full five marks are awarded for correctly labelling all parts.

b) A mark is awarded for correctly describing that an egg is released on day 14. It would have been better to use the correct term ovum in place of egg. A second mark is awarded for saying this is ovulation. A final mark is awarded for saying between days 14 and 28 the lining of the uterus remains thick. This is in preparation for the fertilised ovum to embed and develop into a baby.

c) Three marks are awarded for explaining the reproductive strategies of animals, plants and microorganisms. A further mark is awarded for explaining that sexual reproduction has two parents and genetically different offspring. The final mark is awarded for stating that asexual reproduction has one parent and the offspring are genetically identical.

d) The answer is awarded a mark for stating that substances move from the mother to the baby through the placenta and umbilical cord. A second mark is awarded for giving examples of harmful substances. A final mark is awarded for explaining the baby might die. Additional marks would have been awarded for saying that the baby might have been born prematurely or with a low weight.

a) Label the parts of the female reproductive system. *(5 marks)*

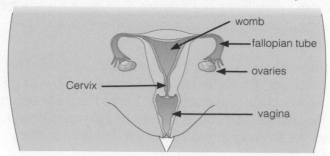

b) Describe what happens in days 14 to 28 of the menstrual cycle. *(3 marks)*

An egg is released on day 14. This is called ovulation. From this point onwards the lining of the uterus remains thick.

c) Explain the difference between sexual and asexual reproduction and in what organisms you might find these processes. *(5 marks)*

Most animals reproduce sexually. Plants can reproduce sexually and asexually. Microorganisms reproduce asexually. Sexual reproduction has two parents and makes genetically different offspring. Asexual reproduction has one parent and makes genetically identical offspring.

d) Explain why smoking and drinking are not recommended for pregnant women. *(3 marks)*

The placenta and the umbilical cord allow substances from the mother's blood into the baby. If the mother drinks then alcohol can go into the baby. If the mother smokes then nicotine can go into the baby. This means the baby might die.

Practice questions

1. This question is about the male and female reproductive systems in humans.

   a)

   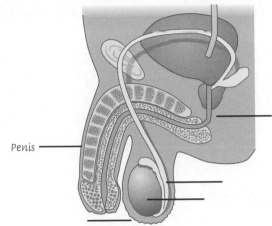

   Penis

   State the names of the parts of the male reproductive system. The first one has been completed for you. *(5 marks)*

   b) Describe the two functions of the penis. *(2 marks)*

   c) Describe the function of the sperm duct. *(1 mark)*

   d)  Explain why millions of sperm are ejaculated into the woman's vagina from a man's penis during sexual intercourse.

   (2 marks)

   e) Explain why sperm are likely to have more mitochondria than ova. *(4 marks)*

2. This question is about conceiving and carrying a baby in humans.

   a) State the name of a fertilised ovum. *(1 mark)*

   b) Describe how a baby is born. *(2 marks)*

   c)  This is a model for the way in which a human embryo is protected. Explain what each part of the model represents.

   (3 marks)

   d) Explain how a baby receives oxygen and food. State the name of the organ which provides this in your answer. *(3 marks)*

3.   This question is about reproduction in other animals.

   a) State the groups of vertebrates that reproduce externally.   *(2 marks)*

   b) Describe how birds reproduce.   *(4 marks)*

   c)

Humans are mammals. Describe one unique way in which mammals care for their young.

(1 mark)

   d) Explain why mammals, reptiles and birds lay fewer ova.   *(2 marks)*

4.   This question is about reproduction in plants.

   a) The flower is the reproductive organ of many plants. The carpel is the female part of the flower. State the names of the parts of the carpel.   *(3 marks)*

   b) Describe the two ways in which plants can be pollinated.   *(2 marks)*

   c) Describe what happens when a pollen grain lands on a stigma.   *(4 marks)*

   d) Describe two ways in which seeds can be dispersed.   *(2 marks)*

   e) Explain why scientists are worried about the reducing numbers of bees found in recent years.   *(2 marks)*

After completing this section you should be able to:
- state the equation for photosynthesis and explain the adaptations of plants to maximise this process
- state the equations for aerobic and anaerobic respiration in animals and micro-organisms and explain the importance of these processes.

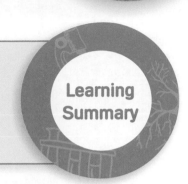

Learning Summary

5.1 Photosynthesis and plant structures

🎧 **15**

Photosynthesis is a chemical reaction which occurs in the green or red parts of plants (leaves, shoots and some stems) that grow on land and in water. Most plants are green because of the pigment **chlorophyll**, which occurs in chloroplasts in their cells. Photosynthesis only occurs in chloroplasts. This reaction allows plants to make their own food (glucose).

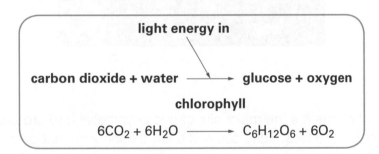

light energy in

carbon dioxide + water ⟶ glucose + oxygen

chlorophyll

$$6CO_2 + 6H_2O \longrightarrow C_6H_{12}O_6 + 6O_2$$

Plants take in carbon dioxide through their leaves and water through their roots. They use the sunlight that is absorbed by their leaves to produce their food (glucose) and oxygen as a by-product. This glucose and oxygen supports the majority of other forms of life (including us).

Plants use the glucose they make in photosynthesis for four purposes:

- respiration
- formation of cellulose to grow new cell walls
- stored as starch
- formation of proteins for growth and repair.

Increased photosynthesis occurs when plants have more light, suitable water, the correct temperature and more carbon dioxide.

Make small cards of all the reactants and products in photosynthesis. Mix them up and then arrange them to give the correct equation.

Plants turn gaseous carbon dioxide and liquid water into sugary glucose and gaseous oxygen during photosynthesis.

In the picture below, photosynthesis only occurs in the green parts of leaves that contain chlorophyll in chloroplasts (so not in the white sections of these leaves).

Algae are aquatic plant-like organisms that also photosynthesise and in fact produce more oxygen than all the plants on land.

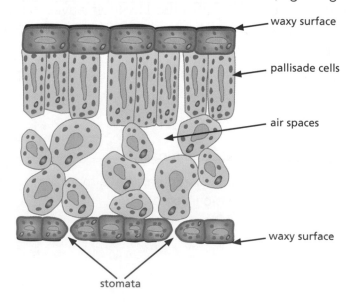

Leaves

The leaves of plants are the major site of photosynthesis and are adapted to maximise this process. They come in a large variety of shapes and sizes to do this. Plants in the shade tend to have larger, darker leaves full of chlorophyll, whilst those in the sun often have smaller, lighter green leaves.

Create a comic strip summarising the importance of photosynthesis to all living organisms, both in providing organic compounds and taking in carbon dioxide from the atmosphere.

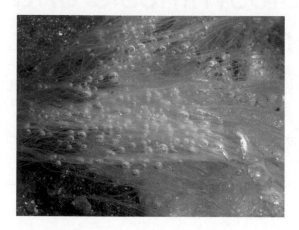

waxy surface

pallisade cells

air spaces

waxy surface

stomata

The adaptations of leaves

Wide	To maximise the amount of sunlight hitting them
Thin	To allow gases to diffuse into all cells easily
Lots of air spaces	To maximise the diffusion of carbon dioxide into and oxygen and water vapour out of the leaf
Tiny holes called **stomata**	To control the volume of gases that can diffuse into and out from the leaf
Waxy surface	To minimise the loss of water vapour from the leaf's surface
Lots of chloroplasts in **palisade cells**	To maximise the rate of photosynthesis

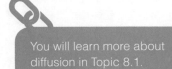

You will learn more about diffusion in Topic 8.1.

Deciduous and evergreen plants

Some plants have evolved to retain their leaves throughout the year. These are called **evergreen** plants. They:

- usually grow nearer the equator where there is lots of sunshine all year round
- or have pine needles to survive cold, dark winters
- or have spines like cacti to survive in hot, dry places.

You will learn more about evolution in Topic 7.1.

Others, called **deciduous** plants, drop their leaves in the winter. These usually live away from the equator where there are shorter, duller days in winter. This means there is not enough light to make it worthwhile for these plants to keep their leaves in winter. They drop their leaves and slow their growth (like hibernation) at this time. This also allows plants to save water.

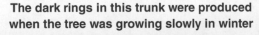

The dark rings in this trunk were produced when the tree was growing slowly in winter

Walk around your garden or local park and look at the leaves of different plants. Can you see patterns in where the plants grew and their leaf size, shape and colour? You could take photographs and use the internet to identify them.

Roots and water transport

Understanding root hair cells in Topic 1.1 will help you here.

Plants absorb water and dissolved mineral nutrients through **root hair cells** in their roots. These massively increase surface area which means plants can absorb more water.

The root hair of a germinating cabbage seed

The length of all the roots and root hairs in this one rye plant is about 380 miles

Collect some leaves that have fallen from deciduous trees in the autumn. Place them between several sheets of kitchen roll in the middle pages of a thick, heavy book for several weeks, to press them flat. Ask an adult to help you make a strong solution of washing powder in a bowl in the sink. Put the leaves into the solution and leave them until the flesh starts to fall away. You could brush them gently with an old toothbrush. Rinse them in cold water to reveal the skeleton.

Once absorbed into the root hair cells, water and dissolved mineral salts flow up through special tubes called **xylem**. Water moves through the xylem to the leaves, where it is used for photosynthesis or evaporates out of the stomata. This evaporation pulls the water from the roots to the leaves and is called **transpiration**. Without the evaporation of some water, transpiration would stop and no water would reach the leaves.

The glucose made during photosynthesis is transported in another type of tube called **phloem**. This moves glucose to where it is stored as starch, which is often in the roots like potatoes or in fruit.

Progress Check

1. State the word equation for photosynthesis.
2. Describe how plants maximise their uptake of water.
3. Explain what transpiration is and why it is important.

5.2 Aerobic and anaerobic respiration

Aerobic respiration

All living cells in plants, animals and microorganisms (except viruses) release energy from glucose. This is called **respiration**. It usually happens in the presence of oxygen and when it does it is called **aerobic** respiration:

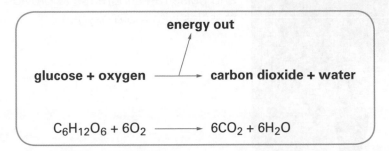

At first look, this equation seems the opposite of photosynthesis earlier in this chapter. But it isn't. Energy from the Sun is required for plants to photosynthesise. So the arrow shows energy going into that reaction. But, in respiration, energy is released from glucose and so here the arrow shows energy leaving the reaction. Organisms use this energy for the other six life processes. The seven life processes can be remembered as MRS GREN:

- movement
- reproduction
- sensitivity
- growth
- respiration
- excretion
- nutrition.

> Make small cards of all the reactants and products in both aerobic and anaerobic respiration. Mix them up and then arrange them to give the correct equations for both aerobic respiration and anaerobic respiration.

Anaerobic respiration in humans

Respiration can also occur without oxygen. This is called **anaerobic** respiration:

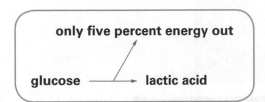

This usually happens when you are exercising vigorously for a long period. Significantly, this process only releases about five percent of the energy of aerobic respiration. The remainder of the energy is stored in the lactic acid. This builds up in your muscles, making them feel 'rubbery', and leads to cramp.

When you finish exercising it takes a few minutes to catch your breath. During this time you breathe quickly and deeply to replenish the oxygen used up in your exercise. This is called **excess post-exercise oxygen consumption** (EPOC). This oxygen reacts with the lactic acid to release the rest of the energy and relieve the cramp.

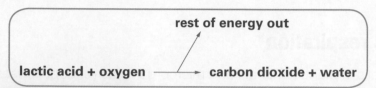

rest of energy out

lactic acid + oxygen ⟶ carbon dioxide + water

During strenuous exercise your breathing and pulse rates increase to provide your muscles with more oxygen

If you use all this oxygen you can only respire anaerobically which leads to cramp

After a short period of rest your body will replenish your oxygen levels and you can return to aerobic respiration

Anaerobic respiration in microorganisms

Anaerobic respiration in some microorganisms is not the same as in humans and other animals. In the fungus yeast and some bacteria:

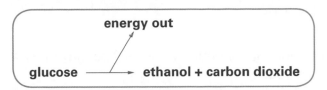

This process is called **fermentation**. The microorganisms complete this process to release the energy needed for them to grow. Ethanol and carbon dioxide are waste products. Ethanol is a type of alcohol and yeast is used to ferment plant sugars to make this alcohol. Beer is made from fermented cereal grains and wine from fermented grapes. Yeast is also used to make bread and the carbon dioxide produced gives us the bubbles in our bread. Bread does not contain alcohol because we do not leave the yeast long enough to form it before we bake the bread and kill the yeast.

Place a few grams of baking yeast into a narrow vase or similar-shaped container. Add a few grams of sugar and enough warm water to make a runny solution. Stretch a balloon over the top of the vase and time how long it takes for the balloon to expand. What gas is causing this? Can you change the mass of yeast or sugar to make the balloon expand faster?

Yeast is a single-celled fungus used to make bread and beer

Make small cards of all the reactants and products in anaerobic respiration in both humans and bacteria. Mix them up and then arrange them to give the correct equations.

Social and economic implications: Yeast is an economically important organism. We use it to make bread and beer.

1. State the word equation for respiration.
2. Describe what a build-up of lactic acid would feel like.
3. Explain how fermentation is different from respiration.

Progress Check

Worked questions

a) State the two products of photosynthesis. *(2 marks)*

Oxygen and glucose.

b) Describe an experiment in which you investigate how light intensity affects the rate of photosynthesis. You may draw a diagram to help you. *(5 marks)*

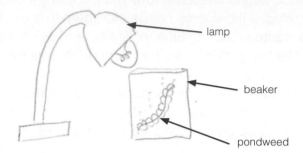

Set up the equipment like the diagram. Move the lamp close to the beaker and count the number of bubbles given off. Then move the lamp progressively further away and repeat the counting.

c) Explain why cells respire. *(2 marks)*

This process releases the energy from glucose so our cells can complete the other six life processes.

d) Yeast is a single-celled fungus. Explain why it is an economically important organism. *(2 marks)*

We use it to make bread and beer. Both of these are sold in lots of shops.

e) Explain the similarities and differences between photosynthesis and respiration. *(4 marks)*

Photosynthesis uses carbon dioxide and water to make glucose and oxygen. Respiration uses glucose and oxygen to make carbon dioxide and water. Energy in the form of sunlight is required for photosynthesis. Energy is released during respiration.

Practice questions

1. This question is about photosynthesis.

    a) State the cell component that is necessary for photosynthesis to occur. *(1 mark)*

    b) State the equation for photosynthesis. Label the reactants and the products. *(4 marks)*

    c) Describe the four uses of glucose for plants. *(4 marks)*

    d) Describe the factors that affect the rate of photosynthesis. *(4 marks)*

    e) Explain why more carbon dioxide would be present in winter in the northern hemisphere when compared with summer. *(3 marks)*

2. This question is about plant structures.

    a) **A plant cell from a leaf**

    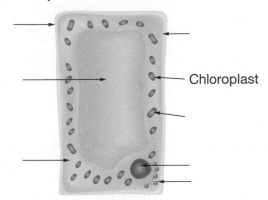

    Chloroplast

    State the cell components of this leaf cell. The first one has been completed for you. *(6 marks)*

    b) State the function of the cell wall and what it is made from. *(2 marks)*

    c)
    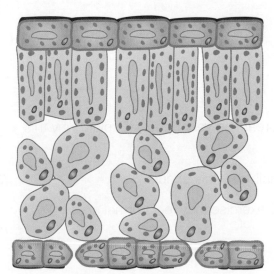

    State the names of the parts of the leaf shown by the arrows. *(5 marks)*

    d) Describe how roots are adapted to absorb water. *(2 marks)*

    e) Explain the process of transpiration. *(2 marks)*

3. This question is about aerobic respiration.

   a) State the cell component in which aerobic respiration occurs. *(1 mark)*

   b) State the word equation for aerobic respiration. Label the reactants and the products. *(4 marks)*

   c) State three uses of the energy released from respiration. *(3 marks)*

   d) Explain why roots of plants are often white. *(4 marks)*

   e)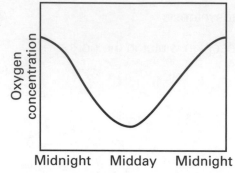

   Describe this graph. *(2 marks)*

   f) Explain the pattern on the graph. *(6 marks)*

4. This question is about anaerobic respiration.

   a) State the equation for anaerobic respiration in microorganisms. Label the reactant and the products. *(4 marks)*

   b) State the equation for anaerobic respiration in humans. Label the reactant and the product. *(4 marks)*

   c)

   At the end of a long race this person is breathing very deeply. Explain why this happens. *(3 marks)*

   d) This person may suffer from cramp. Explain why this occurs. *(3 marks)*

   e) Ten minutes after they have finished their exercise they feel normal again. Explain why this occurs. *(3 marks)*

After completing this chapter you should be able to:
- describe feeding relationships within an ecosystem and explain how organisms depend upon each other
- explain the effects of air and water pollution including pesticides.

Learning Summary

6.1 Relationships within ecosystems

 17

Feeding relationships

A food chain shows the flow of energy between different organisms. The arrows represent the movement of energy. For example:

Grain → Dormouse → Owl

Lots of food chains exist in an **ecosystem** and are shown in a food web:

A food web found in the United Kingdom

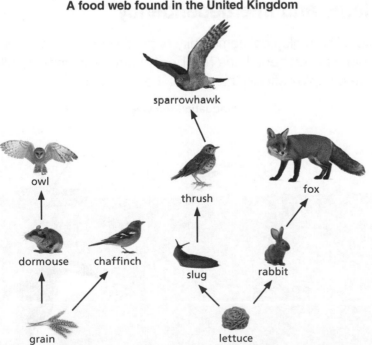

sparrowhawk

owl

thrush

fox

dormouse   chaffinch

slug   rabbit

grain   lettuce

Food chains always start with a **producer**. These are almost always plants, which obtain the energy that is passed along the chain by photosynthesis. Any animal above a producer in a food chain is called a **consumer**. Each stage in a food chain is called a trophic level. Energy is lost at each **trophic level** as the organisms use it to complete the life processes.

You will learn more about photosynthesis and the life processes in Topic 5.1.

Make small cards of each of the organisms in the food web on this page. You can either draw a picture of the organism or just write its name on the card. Mix the cards up and without looking in this book try to arrange the cards to show the correct food web.

The only ecosystems in the world that do not start with plants are found on the bottom of the oceans beside hot volcanic hydrothermal vents. Here, bacteria feed directly on the chemicals released from the vents and the bacteria are eaten in turn by all the other organisms. Life exists here in complete darkness.

Life exists on the ocean floor near hydrothermal vents

Ecosystems and interdependency

An ecosystem is a biological community of organisms that interact with each other and their environment. If the numbers of one organism are altered then this can have complex effects throughout the ecosystem.

A aquatic food web

For example, if the shrimps are all killed as a result of pollution, the seagulls will eat more fish, reducing their numbers.

This example shows that organisms in an ecosystem depend upon each other for survival. We call this **interdependency**. Sometimes small changes to an ecosystem can have devastating and unknown consequences. These changes are sadly often as a result of human interference.

Cutting down trees in the rainforest reduces their numbers but also those of many other organisms that depend upon them

Remove one of the cards from the previous activity to model the death of an organism from your food web. Can you now explain what effects this will have on all the other organisms in your web?

Bioaccumulation of pesticides

Pesticides are chemicals used by farmers to kill pests and so increase their yield of crops. Some pesticides are not excreted by the organisms in food webs and so increase in concentration at each trophic level. This is called **bioaccumulation**.

The particle DDT was a pesticide widely used in World War II to kill mosquitos to stop soldiers catching malaria. DDT was washed into rivers and streams in very low concentrations. Here, it was absorbed by aquatic invertebrates. Because these animals cannot excrete it, all the DDT they absorbed was passed to the small fish that ate them. The concentration of DDT in the small fish was much higher than in the invertebrates. The same thing happened when larger fish ate the smaller fish and finally when birds of prey ate the larger fish. Birds of prey are the top predators in this ecosystem and so had the highest concentration of DDT. The concentration was high enough in these birds to weaken the shells of their eggs, meaning that their chicks died before they hatched.

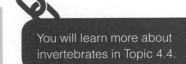

You will learn more about invertebrates in Topic 4.4.

DDT is now banned in many countries such as the UK and USA, but it can still be used in some countries

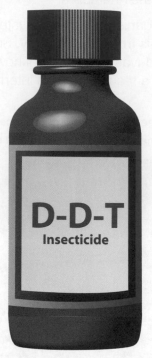

Other examples of chemicals that cause bioaccumulation are mercury metal and PCB chemicals used as coolants in motors. There are reports of pollution of these chemicals leading to bioaccumulation in humans.

DDT accumulates at higher trophic levels, for example in these eagles

Progress Check

1. State the process that organisms at the bottom of almost all food chains complete.
2. Describe why some pesticides bioaccumulate.
3. Explain what interdependency is.

6.2 Pollution

 18

Air pollution

For millions of years the balance of gases in our atmosphere remained relatively stable. Plants used carbon dioxide during photosynthesis and produced oxygen, which plants and animals used to respire, producing carbon dioxide. Unfortunately, in recent years, human activity has released a number of polluting chemicals into the atmosphere, disturbing this balance.

> You will learn more about photosynthesis and respiration in Topics 5.1 and 5.2.

Pollutant	Where from	Consequence
Carbon dioxide	Combustion of fossil fuels	Increases the greenhouse effect which leads to global warming
Carbon monoxide	Combustion of fossil fuels	Carbon monoxide is a poisonous gas
CFCs	Used as coolant liquids in fridges and freezers	Depletes ozone layer which protects us from the Sun's ultraviolet rays
Nitrogen oxides	Combustion of fossil fuels	Contributes to acid rain, smog and breathing difficulties
Sulfur dioxide	Combustion of fossil fuels	Contributes to acid rain

Pollution by CFCs has resulted in a large hole (the red area) in the ozone layer above Antarctica

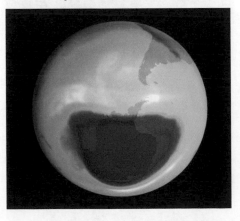

Sulfur dioxide pollution turns rain slightly acidic which can kill trees and chemically weather statues

> You will learn more about the greenhouse effect and global warming in Topic 14.4.

Social and economic implications: Air pollution leads to problems like **global warming**, acid rain and the breakdown of the ozone layer.

Water pollution

Human activity is also polluting fresh and marine water. This affects not only the animals and plants that live in the water but the other animals, including us, that drink it.

Pollutant	Where from	Consequence
Oil	Oil rigs or tankers	Devastating effects on the whole ecosystem
Sewage	Human waste	Spread of disease
Pesticides such as DDT	Overuse or misuse by farmers	Bioaccumulation
Fertilisers	Overuse by farmers	Eutrophication
Metals such as mercury	Factories	Bioaccumulation

Social and economic implications: Water pollution can have terrible consequences for plants and animals, including humans, which rely upon clean water to drink.

Bioindicator species

Bioindicators are organisms that tell us about the level of air or water pollution present in an ecosystem. Some are only present in polluted places whilst others only survive where it is clean. Bloodworms and sludgeworms are found in polluted water but stonefly nymphs are only found in very clean water. All three organisms are bioindicators. Lichens are also bioindicator species. Their presence tells us the air quality is high.

Look at trees or old buildings in your local area. Can you see lots of large lichens? What does this tell you about the quality of air where you live?

With an adult, collect some samples of water from local ponds, stream or lakes. Use a magnifying glass to see if you can see bloodworms, sludgeworms or stonefly nymphs. What does this tell you about the quality of water near you?

Bloodworms

Sludgeworms

Stonefly nymph

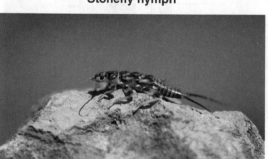

Lichen

Progress Check

1. State the name of the chemicals that make holes in the ozone layer.
2. Describe the effects of acid rain.
3. Explain what the greenhouse effect is.

Worked questions

a) State what organism is present at the bottom of almost all food chains and the relevant process that these organisms complete. *(2 marks)*

Plants. Photosynthesis.

b) State what all other organisms in a food chain are called. *(1 mark)*

Consumers.

c)

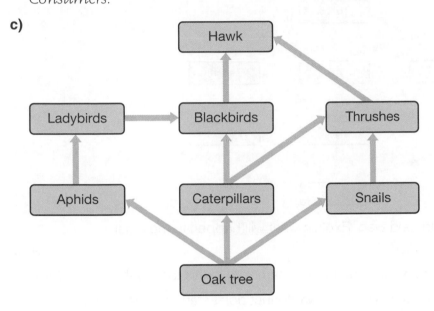

This is a diagram of a woodland food web. Explain what will happen to the other organisms if the ladybirds are killed by pollution. *(4 marks)*

The blackbirds will not have very much to eat. They will eat more caterpillars. The number of aphids will increase because there are no ladybirds to eat them. This means there is less food for caterpillars and snails.

d) Explain how the concentration of pesticides such as DDT changes in food chains. *(2 marks)*

DDT is poisonous. It builds up in food chains because it cannot be excreted.

a) Two marks are awarded for these correct answers. It would be better to write answers in sentences.

b) One mark is awarded for this correct answer.

c) Each sentence in this answer is awarded one mark. With answers about effects of changes to food chains it is very important to set your answer out in a logical way. This will help you get more marks.

d) One mark is awarded for explaining that DDT builds up in food chains. It would have been better to use the word bioaccumulate to describe this process. A further mark is awarded for explaining that pesticides bioaccumulate if they cannot be excreted, so their concentration increases at each trophic level.

Practice questions

1. This question is about feeding relationships.

    a) State what the arrows in a food chain or web show. *(1 mark)*

    b) State what is lost at each trophic level of a food chain and describe what this is used for. *(2 marks)*

    c)
    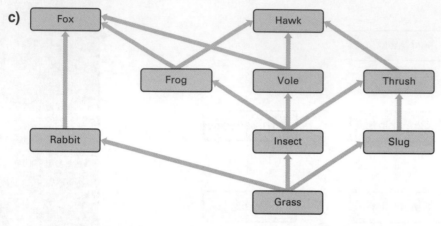

    This is a diagram of a field food web. Explain what will happen to the other organisms if the foxes are killed. *(4 marks)*

    d) Label the producer, a primary consumer and a carnivore in this food web. *(3 marks)*

    e) Describe where the only food chains in the world that don't start with plants are found and explain how life can exist here. *(2 marks)*

2. This question is about pollution.

    a) State the names of two common water pollutants. *(2 marks)*

    b) State the names of two common air pollutants. *(2 marks)*

    c) *(3 marks)*
    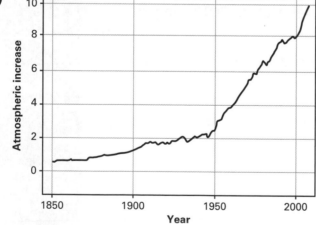

    This is a graph of the concentration of a specific gas in the atmosphere. State the name of the gas and describe why its levels are increasing.

    d) Explain how the greenhouse effect leads to global warming. You can draw a diagram if you would like. *(3 marks)*

    e) Explain how pond dipping could give an indication of how polluted your local stream is. *(4 marks)*

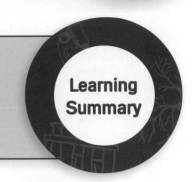

After completing this chapter you should be able to:

- describe how our genetic information is arranged and passed from parent to offspring
- describe the different types of variation
- explain how natural selection leads to evolution.

Learning Summary

7.1 Heredity and genetic diversity

 19

Heredity

You are you because you have inherited some genetic information from both of your parents when your father's sperm fertilised your mother's ovum (egg). As you have grown up, you have developed a personality and have learned many skills and much information. However, your genetic information has not changed. This is unique to you (unless you are an identical twin).

You are the result of sexual reproduction because you are formed from the combination of two sets of genetic information, one from each parent. Many other animals and plants reproduce sexually too. However, some animals, most microorganisms and many plants can reproduce asexually. Here, there is only one parent and the offspring are genetically identical to each other and to their parent. They are called **clones**.

The passing on of genetic information from one generation to the next is called **heredity**.

Identical twins result from sexual reproduction and are genetically identical to each other but not to their parents. A fertilised egg splits into two and grows into twins.

Non-identical twins also result from sexual reproduction but are not genetically identical. Two separate ova are fertilised by two separate sperm, which results in two genetically different people born at the same time. Unlike identical twins, it is possible to have non-identical twins of different sexes.

Many microorganisms reproduce asexually by copying their genetic information before splitting into two in a process called binary fission.

You will learn more about reproduction in Topic 4.2.

Make a family tree using photos of as many of your ancestors as you can. Can you see from whom you have inherited some of your characteristics?

Identical twins

Non-Identical twins

Genetic diversity

The range of genetic differences between organisms within a species is called **genetic diversity**. It is important for the survival of individual populations that genetic diversity of all organisms is maintained. If the organisms aren't genetically diverse, a change to the environment, a new predator or a disease could kill all the organisms, resulting in extinctions. Without genetic diversity there is little or no evolution.

Scientists are very worried about the rate at which humans are destroying habitats like the rainforest. By destroying habitats we are making it much harder for organisms to survive in the first place, but we are also reducing their genetic diversity, which might mean bigger problems in the longer term. Scientists keep genetic material in gene banks to help to maintain genetic diversity. These gene banks commonly store the seeds of plants and cryogenically freeze the sperm and ova of animals.

Some animals such as these pandas are now only present in such small numbers that it is a struggle to maintain their genetic diversity

The Svalbard Global Seed Vault is a plant gene bank in Norway, which is carved into the Arctic permafrost and stores hundreds of thousands of seed samples

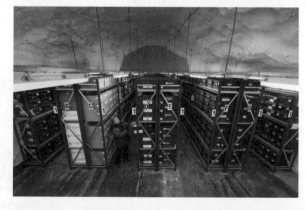

Progress Check

1. State what type of reproduction only has one parent.
2. Describe how non-identical twins are formed.
3. Explain why genetic diversity is important.

7.2 From DNA to cells

The basic unit of **heredity** is deoxyribonucleic acid, or **DNA**, which is found in all living organisms on Earth. DNA is shaped like a twisted rope ladder in a structure called a **double helix**. The rungs of the ladder are pairs of DNA **bases pairs** that are labelled A, T, G and C. The bases A and T pair together and then G and C pair together. There are about three billion base pairs in one copy of your entire DNA (your **genome**).

Every organism resulting from sexual reproduction in every species has a slightly different genome from others in its species. Different species of organisms have genomes with larger differences.

Sections of DNA that are responsible for producing a characteristic like your blood group or eye colour are called **genes**. All your genes are arranged into long, coiled structures called **chromosomes**. There are 23 pairs of chromosomes in most human cells. They come in pairs because you have one set of 23 from your mother and one set from your father.

All cells in your body have 23 pairs of chromosomes except red blood cells (which have no DNA) and your gametes (sperm or ova) which have half this number (23). Other organisms have different numbers of chromosomes. Chimpanzees have 24 pairs whilst many grape vines have 19 pairs.

About 95% of the DNA sequence in a human and a chimpanzee is the same and about 25% of the DNA sequence in a human and a grape vine is the same

Most of your cells contain a nucleus with 23 pairs of chromosomes, split into thousands of genes, made from billions of DNA base pairs

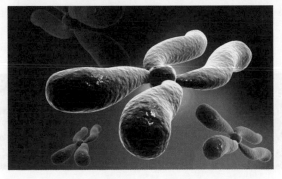

The structure of DNA is a double helix with four possible combinations of base pairs

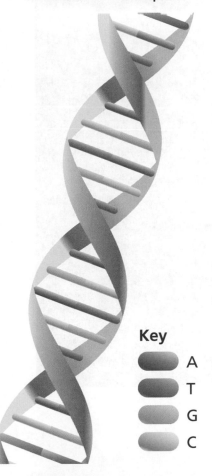

Key

A
T
G
C

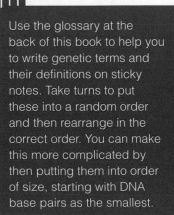

Use the glossary at the back of this book to help you to write genetic terms and their definitions on sticky notes. Take turns to put these into a random order and then rearrange in the correct order. You can make this more complicated by then putting them into order of size, starting with DNA base pairs as the smallest.

Discovering the structure of DNA

You can extract the DNA from peas. You will need an adult to help. Dissolve two grams of salt in 50 cm³ of washing-up liquid in a mug. Mash up 25 grams of peas in a little water in a second mug. Put 10 cm³ of the pea liquid into the first mug. Ask an adult to help you to gently warm the mug and its contents in a pan of water on the stove. Ideally, this would be at about 70°C for 15 minutes. Then put the mug into the freezer for five minutes. Ask an adult to help you very gently pour 50 cm³ of chilled spirit alcohol (vodka works well) into the mug. You will see the DNA as cloudy strands between the pea solution at the bottom and the alcohol that floats on the surface.

In 1953, James Watson and Francis Crick were at Cambridge University. They used X-ray images taken by Rosalind Franklin and Maurice Wilkins, who were working at King's College in London, to come up with a model of DNA. Sadly, Franklin died in 1958, several years before Watson, Crick and Wilkins were awarded the Nobel Prize for their achievement. At this time, the prize could only be awarded to the living and so was not given to Franklin even though her work was instrumental in the discovery.

In 1953, Watson and Crick discovered the structure of DNA

The X-ray image that Franklin and Wilkins took, which helped Watson and Crick discover the structure of DNA

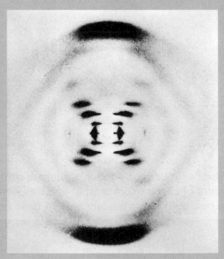

Progress Check

1. State what a gene is.
2. Describe the structure of DNA.
3. Explain why controversy surrounds Rosalind Franklin and the discovery of DNA.

7.3 Variation

There are differences within a species and between different species. For example, all domestic cats are the same species but they can look very different. Differences within a species like this are called **variation**. Smaller groups within species like cats, dogs and horses that have less variation within them are called breeds. In humans we call these groups 'races'.

If these differences are inherited from parents we call this **genetic variation**. Examples in humans include attached or unattached earlobes and whether you can roll your tongue or not. If it results from the conditions in which an organism has lived we call it **environmental variation**. Examples include scars in humans and flower colour in hydrangea plants.

Some variation is caused by both genetic and environmental factors. People can inherit genes that make them taller or broader. This is genetic variation. However, their diet and lifestyle can also affect how tall or broad they are. This is environmental variation. So body shape is often affected by both genetic and environmental variation.

Whether your ears have attached or unattached lobes and whether or not you can roll your tongue are examples of genetic variation

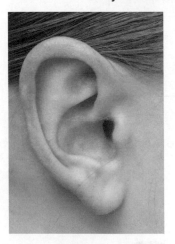
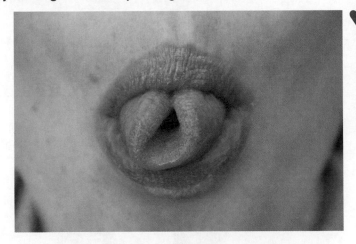

Add the following into the family tree diagram that you made earlier for each of your relatives: eye colour, attached or unattached ear lobes and whether or not they can roll their tongue. Can you spot any patterns?

Hydrangea plants grown in acid soil have blue flowers whilst those grown in alkaline soil have pink flowers. This is not genetic so is an example of environmental variation

Continuous and discontinuous variation

Genetic and environmental variation can be further categorised:

Category of variation	Why	Example in humans	How presented
Continuous	The data collected comes in a range	Height, weight, hand span	Line graph
Discontinuous	The data collected comes in discrete groups	Eye colour, shoe size, blood group	Bar chart

An example of discontinuous variation is the percentages of people in the UK with the four different blood groups (O = 44%, A = 42%, B = 10%, AB = 4%)

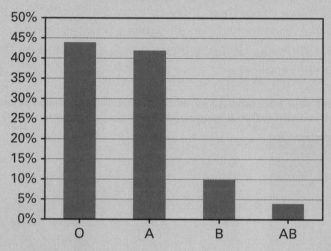

An example of continuous variation is the range of human height

Measure the heights and record the eye colours of a group of your family or friends. Draw two graphs to show the data. You will need to decide what type of graph can best be used to represent each type of data.

Progress Check

1. State the two causes of variation.
2. Describe the differences between continuous and discontinuous variation.

7.4 Natural selection and evolution

The theory of **evolution** explains how the diversity of life we see on Earth today has grown from the first life forms. These were simple single-celled organisms and probably existed over two billion years ago. The steps in evolution are:

1. Variation exists in the populations of all living organisms.

2. Because the numbers of organisms of most populations remain the same, the individuals within them must compete with each other to survive.

3. Variation results in some individuals possessing characteristics which mean that they are better adapted to survive and reproduce than others ('**survival of the fittest**' or '**natural selection**').

4. Heredity means that these advantageous characteristics are passed to their offspring.

5. Over many, many generations, small changes in these characteristics can add up and new species of life are formed.

Those individuals or entire species that are poorly adapted may die out or become **extinct**.

We can see how some organisms have undergone these small changes over time by looking at fossils. These provide evidence for evolution. There are some gaps however, for not all organisms become fossilised and not all fossils have been found. We can also see these small changes in organisms that reproduce much faster than us, like bacteria or small fruit flies.

The fossil record provides evidence for evolution

Evolution often occurs as a result of environmental change. We can see small alterations in the structure of horses over time. As the grassland they lived on became less wet they needed smaller feet to allow them to run faster and avoid predators

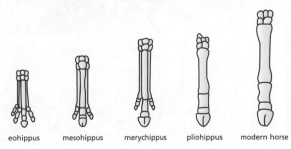

eohippus   mesohippus   merychippus   pliohippus   modern horse

Devising the theory of evolution

Charles Darwin was a famous English scientist who lived in the 19th century. He travelled around the world on a ship called HMS *Beagle*. The ship stopped at a small group of islands in the Pacific Ocean called the Galapagos Islands. Here, Darwin observed similar-looking finches on different islands, which had a range of different beak shapes and sizes. Darwin thought that the birds all belonged to one species that was blown from the South American mainland to the islands in a storm. He reasoned that the small changes that must have existed in the original population of birds allowed them to feed successfully on specific islands. From this thought he developed his theory of evolution by natural selection following the steps on the previous page.

Darwin returned to England and wrote his ideas into a book called *On The Origin of Species by Means of Natural Selection*. But he did not publish his work immediately. At this time, the Church was very powerful and Darwin was worried that his theory disagreed with the Bible. Twenty years later, another scientist called Alfred Russel Wallace came up with the same theory. He wrote to Darwin to explain it. Both scientists jointly published their findings but it is Darwin who is the best remembered today. Darwin was right to be worried, for the Church attacked him and his theory shortly after its publication.

Charles Darwin saw how these finches had developed beaks that were best adapted to their island and from this thought of the theory of evolution

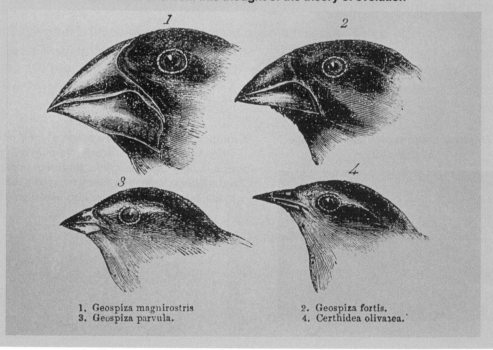

1. Geospiza magnirostris
3. Geospiza parvula.
2. Geospiza fortis.
4. Certhidea olivaɔea.

Imagine that you are Charles Darwin aboard HMS *Beagle* coming back to England. Write a letter to explain your theory of evolution and how you came up with it after seeing the finches on the Galapagos Islands. Stain your letter using a used teabag to make it look old. Ask an adult to help you to burn the edges too. This is easiest if you roll your letter up first.

Progress Check

1. State what must be present in a population for evolution to occur.
2. Describe what survival of the fittest means.
3. Explain why Darwin did not publish his findings immediately.

Worked questions

a) State the unit of heredity. *(1 mark)*

The gene.

b) State the two categories, not causes, of variation. *(2 marks)*

Continuous and discontinuous.

c) Describe the differences between sexual and asexual reproduction. *(4 marks)*

Sexual reproduction involves two parents and asexual reproduction has one parent. The offspring are genetically identical.

d)

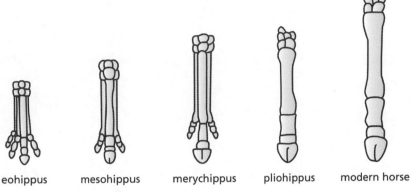

eohippus   mesohippus   merychippus   pliohippus   modern horse

This diagram shows how horses' feet have changed over thousands of years. Explain how this provides evidence for evolution. *(5 marks)*

The steps in evolution are:

1. *Every population has variation.*

2. *There is competition within populations (as their overall numbers remain stable).*

3. *Some organisms will be better adapted to survive and reproduce.*

4. *These adaptations will be passed on to their offspring.*

5. *These small changes eventually form new species.*

So the marshes dried up and become grass. The better adaptation was smaller feet to run from predators. Fossils show horse's feet becoming smaller.

a) One mark is awarded for stating that the gene is the basic unit of heredity.

b) Two marks are awarded for stating the two categories. It is important to not confuse these with the two causes: genetic and environmental.

c) Two marks are awarded for saying that sexual reproduction involves two parents and has genetically different offspring. Two further marks are awarded for saying asexual reproduction involves one parent which produces genetically identical offspring (clones).

d) Full marks are awarded for correctly stating the steps in evolution and for explaining how it relates to horse's feet.

Practice questions

1. This question is about genetic structures.

   a) Put the following terms in order of size starting with the smallest:
      Genome / Genes / DNA base pairs / Chromosomes   *(2 marks)*

   b) State the definition of a genome.   *(1 mark)*

   c) 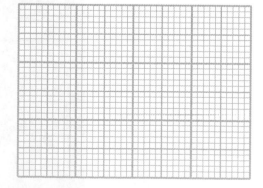 Describe the structure of DNA. One base pair has
      been identified with an A and one with a G. Add the
      other letters to all the remaining bases.   *(4 marks)*

   Key
   A
   G

   d) Explain how the structure of DNA was discovered.   *(3 marks)*

2. This question is about variation.

   a) State the definition of variation.   *(1 mark)*

   b) Describe the ways in which variation arises, giving an example of each in
      your answer.   *(4marks)*

   c) The table shows the results of a survey of the eye colour of your class. State the
      category of variation being investigated here. Draw a graph to show the results
      of this survey.   *(4 marks)*

Eye colour	Number in class
Blue	7
Brown	17
Green	3

d) Describe the results of this graph. *(3 marks)*

e) Explain how and why you would draw a graph showing the results of a survey of the specific heights of people in your class. *(2 marks)*

3. This question is about evolution.

a) State the names of two scientists who independently devised the theory of evolution. *(2 marks)*

b) Describe how this theory was first devised. *(4 marks)*

c) Explain why the theory of evolution was not immediately published. *(1 mark)*

d) Explain why some organisms have become extinct. *(2 marks)*

e)

Before the Industrial Revolution, most trees were pale coloured and so were most moths. The Industrial Revolution produced lots of smoke, which turned many trees black. Several years later, most of the moths were also black. Explain how this provides evidence for evolution. *(6 marks)*

4. This question is about heredity.

a) State the two reproductive cells in plants. *(2 marks)*

b) Describe how microorganisms reproduce. *(1 mark)*

c) Explain why reproductive cells have half the genetic information of a normal body cell. *(2 marks)*

d) Explain why you are similar to your brothers and sisters but not exactly the same. *(2 marks)*

e) Explain why identical twins must be of the same sex but non-identical twins can be different sexes. *(2 marks)*

8 Chemical fundamentals

Learning Summary

After completing this chapter you should be able to:
- explain the typical properties of solids, liquids and gases using ideas about particles
- describe what happens to particles when a substance changes state and use ideas about energy to explain this
- explain the difference between physical changes and chemical reactions
- describe the principle of conservation of mass.

8.1 Solids, liquids and gases

Substances can generally be classified into three groups: solids, liquids and gases. These are called the three states of matter.

The particle model

All substances are made from particles. These particles might be **molecules**, **atoms** or **ions**. By considering the arrangement and movement of these particles, scientists can explain the typical properties of solids, liquids and gases.

Solid	Liquid	Gas
• Particles are touching their neighbours, **so solids are usually dense**	• Particles are mostly touching their neighbours, **so liquids cannot be compressed**	• Particles are widely spaced, **so gases are compressible and have lower densities than solids and liquids**
• Particles are fixed in place by very strong forces, **so solids often have a high melting point**	• Particles are attracted to each other by fairly strong forces, **so liquids have lower melting points than most solids**	• Particles are only attracted to each other by very weak forces, **so gases have very low melting points**
• Particles move only by vibrating, **so solids have a fixed shape**	• Particles can move over each other, **so liquids can take the shape of their container**	• Particles are moving very quickly, **so gases move through space or can take on the shape of the container**

Evidence for the particle model

Scientists develop models to help them explain things that they observe. A model doesn't have to be something you can see and touch. It might be an idea or a way of explaining something in a simple way so that others can understand it.

The particle model is not perfect but it does help us to explain the properties of solids, liquids and gases. It also explains **diffusion** and **Brownian motion**.

Diffusion

When bread is cooking, you can often smell it from another room. This is because the particles that cause the smell are able to travel through the air. This process is called diffusion and it takes place in liquids and gases but not solids. How does diffusion work?

- The smell particles can move between the air particles.
- The air particles bump into the smell particles and push them around.
- The smell particles move, over time, from an area where there are lot of them (high **concentration**) to areas where there were fewer of them (low concentration).

Diffusion occurs in liquids too, but more slowly. Why do you think this is?

The blue liquid diffuses through the water. Eventually, the concentration of the blue liquid will be equal throughout the mixture.

Brownian motion

In 1829, Robert Brown looked through a microscope at grains of pollen in water. He noticed that they moved randomly, changing direction all the time. He could not explain why, and nor could anyone else for over 75 years! In 1905, Albert Einstein explained why the pollen grains moved in the water: they are constantly being hit by water particles, which are far too small to see using a **light microscope**. At any one instant, if a pollen grain is hit more on one side than on the other, it will move.

Make a 3D model of a solid, a liquid and a gas. See if you can make your models in such a way that they help to explain some of the properties of the three states of matter.

Pressure in gases

You will learn more about pressure in gases and in liquids in Topic 16.7.

The particles in a gas are widely spaced and move very quickly. When they bump into the sides of the container, they press against it. The particles are very small and have a very low mass but there are millions and millions of them, so the total effect is a noticeable force, pushing outwards. This is called gas **pressure**.

Several factors affect the pressure of a gas.

The number of gas particles

Fewer particles		More particles	
	The particles collide with the walls of the container less frequently, so the **pressure is low**.		The particles collide with the walls of the container more frequently, so the **pressure is high**.

The volume of the container

Large volume		Small volume	
	The particles collide with the walls with a low frequency, so the **pressure is low**.		When the volume of the container is decreased, the **pressure increases**. This is because the particles collide with the walls more frequently.

The temperature

Low temperature		High temperature	
	The particles move more slowly, so they exert less pressure on the sides of the container.		The particles move faster, so they exert a greater pressure on the sides of the container.

Changes of state

Most pure substances can exist as a solid, a liquid or a gas. A substance can be changed from a solid to a liquid or a gas by changing the temperature.

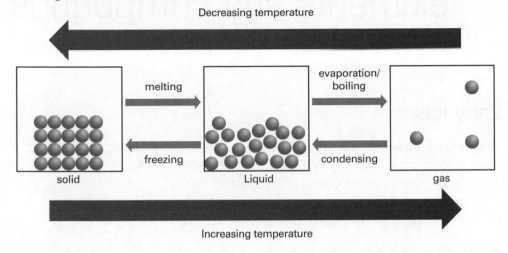

When a solid is heated, it **melts** into a liquid. For example, when you heat chocolate, it turns into liquid chocolate. When it cools again, the chocolate freezes (solidifies).

When a liquid is heated, it turns into a gas. For example, water turns into steam. If this process takes place slowly at a temperature lower than the boiling point, this is called **evaporating**. This forms a vapour. If it takes place quickly at the boiling point, it is called **boiling**. The reverse of this process is called **condensing**, which takes place when a gas cools. When steam condenses, it forms liquid water again.

Sublimation

In some situations, a solid may turn straight into a gas when it is heated. This is called **sublimation**. Solid carbon dioxide is called dry ice and it sublimes at room temperature and pressure. It is possible to turn carbon dioxide gas into a liquid but only under high pressure, which forces the particles closer together to make a liquid.

Water is unusual

The diagrams on this page suggest that, for a pure substance, the solid should be more dense than the liquid and the liquid should be more dense than the gas. Water is unusual because ice is less dense than liquid water, which is why ice floats.

Work with some friends to create a role play that explains to primary school pupils what happens to the particles in a solid when it is heated enough for it to melt and then boil. Remember to include details of the arrangement of particles as well as their movement.

1. When someone in the changing rooms sprays deodorant, after a few minutes you can smell it by the door. Explain why.
2. After a car journey, the tyres on a car feel hot. Would you expect the pressure of the air inside the tyre to be higher or lower than before the journey? Explain your answer.
3. Describe the changes that will take place if a lump of solid copper is heated, slowly, to a very high temperature.
4. Suggest why energy must be supplied to a liquid to turn it into a gas. Use ideas about the attractive forces between particles in your answer.

Progress Check

8.2 Atoms and molecules; elements and compounds

The smallest building blocks that every substance is made from are called atoms.

Early ideas

The idea of the atom was first proposed by an ancient Greek philosopher called Democritus. He was not a scientist because he did not test his ideas using experiments or models. He suggested that...
- All matter is made from atoms.
- Atoms cannot be divided into smaller particles.

Dalton's atomic model

In the early 1800s, John Dalton suggested some improvements to the atomic model...
- Each chemical element is made from only one type of atom.
- Atoms cannot be made or destroyed in chemical reactions or physical changes.
- Atoms of one element can combine with atoms of another element in fixed ratios to form a compound with a definite chemical formula.

Elements and compounds

An **element** is a pure substance which is made from only one type of atom. The elements are found listed on the periodic table, along with their unique symbols. For example, the symbol for oxygen is O, sodium is Na and helium is He. There are around 100 elements that are found naturally.

Elements combine to form compounds. **Compounds** are pure substances that have different **properties** from those of the elements from which they are made. Hydrogen is a flammable gas. Oxygen is a reactive gas. These two elements react to form water, which is very different from the hydrogen and oxygen. Compounds have a **chemical formula** that tells you how many of each type of atom they contain. H_2O tells you that there are two hydrogen atoms for every oxygen atom. Compounds made from ions form giant structures.

Atoms and molecules

Some substances are made from molecules. These are clusters of atoms that are joined by very strong chemical bonds, which are broken only by chemical reactions, not physical changes. For example, a molecule of ammonia, NH_3, is made from one nitrogen atom and three hydrogen atoms. Remember that not all substances are made from molecules. Some compounds are made from ions instead and you will learn about these for your GCSEs.

Atomic structure

Scientists continued to study the atom after Dalton proposed his theory in the early 1800s. They did experiments to find out whether the atom was made up of smaller particles. The evidence from these experiments allowed them to conclude that atoms were made up of smaller particles, called protons, electrons and neutrons. Recent research, in particle accelerators like the Large Hadron Collider at CERN, has discovered more tiny particles that are smaller than atoms. You may have heard about the Higgs boson, for example. This is a very active field of research.

An atom is made up from protons, electrons and neutrons. It is the number of these particles present in an atom (specifically the number of protons) that makes an atom of one element different from an atom of another element. Here is a diagram of a lithium atom.

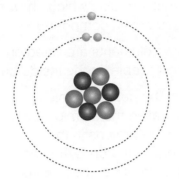

● proton

● neutron

● electron

The **protons** and **neutrons** are found in the **nucleus**, which is at the centre of the atoms and where (almost) all of the mass of the atom is found. The electrons orbit the nucleus in shells. Each shell has a maximum number of electrons. The first (innermost) shell can hold two electrons, so the third electron in a lithium atom must go into the next shell.

Other atoms are different. Here is a diagram of a boron atom.

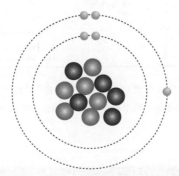

Boron atoms contain five protons. This means that they are element number 5 on the periodic table. We say that they have an **atomic number** of 5.

Here is a summary of the differences between protons, electrons and neutrons.

	Name of sub-atomic particle	Mass	Charge	Where is it found?
●	Proton	1	+ (positive)	In the nucleus
●	Neutron	1	No charge (neutral)	In the nucleus
●	Electron	Almost zero	− (negative)	Orbiting the nucleus in shells

Chemical symbols and the periodic table

In Topic 12.2, you will learn about how the periodic table was developed and how we can use it to classify elements and understand patterns in their reactions.

Chemists use symbols to help them to communicate effectively. This is because…
- The same symbols are used no matter what language a chemist speaks and writes in.
- It saves time.
- Symbols can be made into formulae which show how many of each type of atom are bonded together in a compound.

All the symbols for the known elements are listed on the periodic table. There is a copy of the periodic table on the inside back cover.

Here is a square taken from the periodic table. The smaller number (26) is called the atomic number, which is the number of protons in the nucleus of an atom of this element. Iron is element 26 in the periodic table. The larger number (56) is the mass number and is the number of protons plus neutrons.

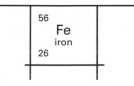

Sometimes the symbols used seem obvious. For example, H is hydrogen. Sometimes, the symbol comes from the Latin name for the element. For example, Fe stands for *ferrum*. The first letter of the element symbol is always a capital.

Chemical formulae

Use coloured card to cut out circular atoms of carbon (black), oxygen (red) and hydrogen (white, smaller than the other atoms). Use them to model the following molecules: CH_4, CH_4O, C_2H_4, C_2H_6O.

A chemical formula shows how many of each type of atom are bonded together. For example, the most common type of iron oxide has a formula Fe_2O_3.

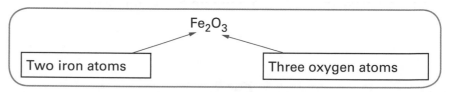

Fe_2O_3

| Two iron atoms | Three oxygen atoms |

It is sometimes helpful to write out a formula like this: FeFeOOO. Remember that some elements have two letters in their symbols, so don't write out FeeOOO!

Progress Check

1. Write down the chemical symbols for the following elements: sulfur, magnesium, neon, nitrogen, potassium and aluminium.
2. What atoms are present in a molecule of ethane, C_2H_6? State how many of each type there are.
3. Why will you not find water listed in the periodic table?

8.3 Chemical reactions

In a **chemical reaction**, new chemicals are made. Examples of chemical changes include burning a fuel, rusting, respiration and photosynthesis. A chemical change can usually be identified because...

- They are usually very hard to reverse.
- There is a large energy change (energy given out or taken in).
- There might be a colour change.
- There might be evidence of a new chemical being made (e.g. bubbles).

Remember that dissolving is not a chemical reaction and nor are any of the changes of state (e.g. melting, boiling). It is very easy to reverse these processes, so they are called **physical changes** instead.

What happens to atoms in a chemical reaction?

In a chemical reaction, atoms are rearranged. This means that chemical bonds between atoms are broken and then new chemical bonds are made. We can show this using particle diagrams. For example, here is a particle diagram to represent the reaction of hydrogen molecules with oxygen molecules to form water molecules. You can see that the chemical bonds between the hydrogen atoms have been broken. The bond between the oxygen atoms has been broken too. New bonds have been made in the water molecules.

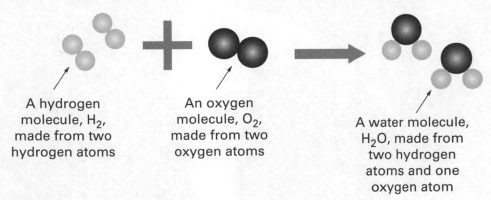

A hydrogen molecule, H_2, made from two hydrogen atoms

An oxygen molecule, O_2, made from two oxygen atoms

A water molecule, H_2O, made from two hydrogen atoms and one oxygen atom

The thermite reaction

Conservation of mass

In a chemical reaction, atoms are rearranged. They are never created and they are never destroyed. This means that all the atoms present at the start of the reaction are still present at the end. This is called the **law of conservation of mass**. We say that mass is conserved in a chemical reaction.

For example, in one chemical reaction, iron oxide reacts with aluminium to make aluminium oxide and iron. We can measure the mass of the chemicals before and after the reaction...

iron oxide + **aluminium** → **aluminium oxide** + **iron**

(160 g)   (54 g)   (102 g)   (112 g)

If you add up the masses of the chemicals that reacted together, you get 160 + 54 = 214 g. If you add up the masses of the chemicals that were produced in this reaction, you get 102 + 112 = 214 g. You can see that mass is conserved.

Mass is also conserved in physical changes. If you melt 5 g of ice, you will have 5 g of water.

Reactions that seem to get lighter or heavier

When you burn a wooden splint, it seems to get lighter. This does not mean that atoms have been destroyed! The chemicals produced in some reactions are gases and they escape into the air, so it is easy to forget them. If you collected the gases and added their mass to the mass of the ash, you would find that mass is conserved in this reaction too.

When you burn magnesium, it seems to get heavier. This extra mass has come from the oxygen atoms that were in the air but are now chemically bonded to the magnesium atoms.

Make a model of a seesaw and some atoms that allows you to demonstrate that because atoms are rearranged in chemical reactions (but not created or destroyed) that the mass is conserved.

Progress Check

1. Which of these are chemical reactions and which are physical changes?
   a) Cooking an egg
   b) Dissolving salt in water
   c) Burning some paper
   d) Freezing water
2. Why is mass always conserved in chemical reactions and physical changes?

Worked questions

a) Substances can be described as solid, liquid or gas. Draw diagrams in the boxes below to show how the particles are arranged in a solid, a liquid and a gas. Use a **minimum** of six particles in each box. *(3 marks)*

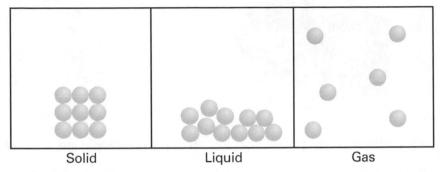

| Solid | Liquid | Gas |

b) State two properties of a gas and explain them using ideas about particles. *(4 marks)*

Gases can are compressible, because they are made from widely spaced particles.

Gases also have a low density, which is also because the particles are widely spaced.

c) Describe how the movement of particles in a gas changes when the temperature is increased and the effect that this has on the pressure. *(2 marks)*

The particles move faster when a gas is heated, so the pressure increases.

d) If the volume of the container is reduced at constant temperature, why will the gas eventually become a liquid? *(2 marks)*

Reducing the volume of the container will increase the pressure. This pushes the particles closer together until they are touching. This means the gas will have turned into a liquid.

a) Remember to show the right number of particles asked for in the question. The solid particles must be regularly arranged and mostly touching. The liquid particles must be mostly touching but have a random arrangement. The gas particles must be widely spaced and have a random arrangement.

b) The question says **state** which means that you can give a simple answer. In this case, two marks are awarded for correctly listing two properties of gases (can be squashed and low density). The question also says **explain** which requires a detailed scientific answer, often linked with words like **because**.

c) It is important to use the correct scientific words in your answers. Sometimes, students say that when the temperature increases, the particles 'move more' which will not earn the mark. But saying 'move faster' is correct.

d) The first mark is awarded because the student has correctly linked decreasing volume to increasing pressure. The second mark is awarded for linking this process with the particles eventually touching each other.

Practice questions

1. This question is about water.

a) Solid water is called ice. Describe the arrangement and movement of particles in ice. *(3 marks)*

b) What name is given to the process when ice turns into liquid water? *(1 mark)*

c) Explain why ice turns into liquid water when it is heated. In your answer, use ideas about energy and the attractive forces between molecules of water. *(2 marks)*

d) Ice floats on water, which means that ice is **less dense** than liquid water. Use the particle model of matter to explain why this fact is unusual. *(2 marks)*

e) Gaseous water is called steam. Describe the arrangement and movement of the water molecules in steam. *(2 marks)*

2. This question is about evidence for the particle model.

a)

Smoke is made from tiny particles of solid ash that are suspended in the air. These ash particles are too small to be seen with the eye but you can see them using a light microscope. They are much larger than the molecules of the gases that make up the air. When you look at smoke through a light microscope, the tiny ash particles move in random directions. This is called 'Brownian motion'. Explain what causes Brownian motion and how this provided scientists with evidence that gases are made from particles. *(4 marks)*

b) Air freshener is often sprayed in a room to make the room smell nice. It is only necessary to spray it in one corner of the room and eventually you can smell it everywhere. Explain why this happens and how this gives us evidence for the particle model. *(3 marks)*

c) The same process that allows the air freshener to be smelled everywhere in the room also occurs in liquids but it is much slower. Explain why. *(2 marks)*

3.   Look at the particle diagrams below. Each circle represents an atom. Different elements are shown in different shades. Answer the questions that follow.

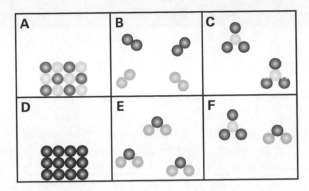

a) (i) Which of these diagrams shows a pure element?

    (ii) What state is this element in?   (2 marks)

b) Which diagrams show a pure gaseous compound?   (1 mark)

c) Which diagram shows a mixture of gaseous elements?   (1 mark)

d) Which diagram could represent H_2O?   (1 mark)

e) Which diagram could represent FeO?   (1 mark)

f) Which diagram shows a substance that is **not** made from molecules? (1 mark)

4.   This question is about chemical reactions and physical changes in the kitchen.

a) Explain why cooking an egg involves chemical reactions and not physical changes?   (3 marks)

b) Sometimes, eggs are cooked by placing them in boiling water. Explain why boiling water is a physical change and not a chemical reaction.   (2 marks)

c) Suggest what happens to the mass of an egg when it is boiled in water. Explain your answer by explaining what happens to atoms in a chemical reaction.   (2 marks)

After completing this chapter you should be able to:
• state some examples of pure and impure substances
• describe the process of dissolving
• select an appropriate method of separating mixtures of substances.

 9.1 # Pure substances and mixtures

A sample of pure gold

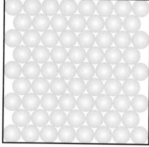

A substance is **pure** if it contains only one chemical **compound** or **element**. For example, if a sample of gold (an element) is pure, then it contains only gold **atoms**. If a sample of water (a compound) is pure, then it contains only water **molecules**.

If a substance contains more than one chemical, then it is described as a **mixture**. The second chemical may be present in very small amounts. In this case, it is described as an **impurity**.

A sample of gold that is impure. The grey atoms are the impurities

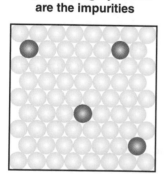

Common mixtures

Most of the substances in our everyday lives are actually mixtures. Even when we think they are pure substances, there are likely to be impurities in them. Tap water contains around 300 particles of impurities for every one million water molecules.

The air we breathe is a mixture of gases. Some are elements, including nitrogen (N_2) and oxygen (O_2). Others are compounds, including carbon dioxide (CO_2).

Impurities can sometimes be dangerous

Sometimes we need things to be very pure. For example, the metal used inside a mobile phone or computer must be very pure to allow it to conduct electricity effectively. And when you take a medicine, you want to be sure that there are no harmful impurities present.

Progress Check

1. Which of the pictures shows a pure substance?
2. Is the pure substance made from atoms or molecules?

A

B

C

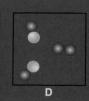
D

9.2 Dissolving

🎧 **27**

When you dissolve some salt into some water, you have made a special type of mixture called a **solution**. Here are the key words you need to know, using the salt solution as an example.

Key word	Definition	Example
Solute	The chemical (usually a solid) that is dissolved into the solvent	In a salt solution, the solute is the salt.
Solvent	The liquid that the solute is dissolved in	In a salt solution, the solvent is the water.
Soluble	Used to describe a solid that can dissolve in a particular solvent	Salt is soluble in water.
Insoluble	Used to describe a solid that cannot dissolve in a particular solvent	Wax is insoluble in water but it is soluble in propanone.

Making solutions

Water is colourless. Copper sulfate is a blue solid. When you dissolve copper sulfate into water, you get a blue solution. The more solute you add, the more intense the colour gets.

Solid copper sulfate

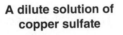
A dilute solution of copper sulfate

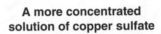
A more concentrated solution of copper sulfate

When a white solute dissolves into a colourless solvent (like salt dissolving into water), you get a colourless solution. The crystals of solute break into particles that are too small to see.

Make a model to explain to a younger student what happens when crystals of copper sulfate dissolve in water.

1. When sugar is dissolved into water, what is the solvent and what is the solute?
2. When you make up a solution of copper sulfate, why do the crystals seem to disappear?

Progress Check

9.3 Separating mixtures

When elements **react** to produce a compound, it is very hard to break the **strong chemical bonds** between the atoms. However, when substances are **mixed** together, it is usually easier to separate them.

The technique you use depends on the properties of the substances that are mixed.

For example, iron is magnetic. Sulfur is not. If iron filings are mixed with sulfur, it is easy to separate them using a magnet to attract the iron.

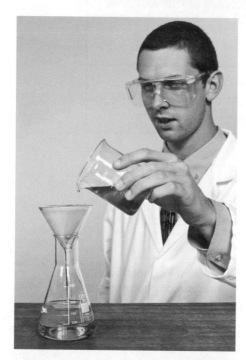

Filtration

If one of the substances is insoluble, it can be separated from a liquid or a solution of another substance by filtration.

1. Fold your filter paper and then place it in a filter funnel.

2. Place your filter funnel into a conical flask.

3. Pour the mixture into the filter paper.

4. The liquid (or solution) will pass through the filter paper. It collects in the flask and is called the **filtrate**.

5. The insoluble solid will not pass through, and remains as the **residue**.

Evaporation

Evaporation can be used to separate the solute from a solution when the solvent is not needed. For example, to get the sodium chloride back from a solution of sodium chloride, you would evaporate the water.

1. Pour your solution into an evaporating basin.

2. Place the evaporating basin on a wire gauze which is resting on a tripod.

3. Use a Bunsen burner to heat the solution gently until it boils.

4. Evaporate half the solvent.

5. Turn off the Bunsen burner and leave the rest of the solvent to evaporate slowly in a warm place.

6. You will be left with crystals of sodium chloride.

Distillation

Distillation can be used to separate two liquids that have different boiling points or it can be used to separate a solvent from a solute when you want to keep the solvent. This is how distilled water is made in science labs, so it is much more pure than tap water.

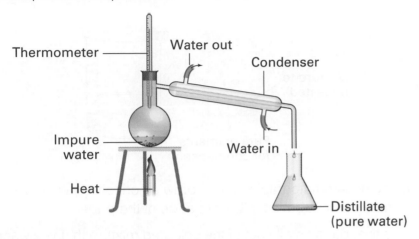

1. Place the mixture to be separated into a flask.

2. Connect a condenser to the side of the flask. The cold water always enters at the bottom of the condenser.

3. Place a flask for the pure distillate at the lower end of the condenser.

4. Heat the mixture until the liquid boils. As it passes into the condenser, it is cooled by the water running around the outside.

Fractional distillation

If the mixture contains a number of liquids that have different boiling points, you can use the same apparatus as for simple distillation. After the first liquid has been distilled off, change the conical flask and heat the mixture in the round-bottomed flask until it reaches the boiling point of the next liquid. You will then be able to condense this gas and collect the second distillate.

Social and economic considerations: Fractional distillation is used to extract useful chemicals like petrol and diesel from crude oil. Without this process, we would have no fuels for transport and no chemicals from which to make plastics!

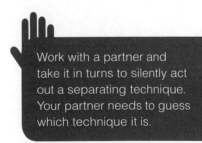

Work with a partner and take it in turns to silently act out a separating technique. Your partner needs to guess which technique it is.

A refinery for producing chemical products

Chromatography

Chromatography is used to separate different solutes that are dissolved in the same solvent. It is an easy way to separate food colour pigments but it is also one of the techniques used to test the urine of sportsmen and women to make sure that they are not taking illegal drugs to boost their performance.

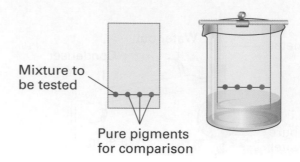

Mixture to be tested

Pure pigments for comparison

1. On a piece of chromatography paper, draw a horizontal pencil line approximately 2 cm above the bottom edge of the paper.

2. On to this line, put a small spot of the coloured mixture that you want to separate. Make this as small but as intense as you can, by reapplying it several times.

3. Put spots of other pure pigments in different places, spaced out further along the line.

4. Pour some water into a large beaker so that it is 1 cm deep.

5. Place the piece of chromatography paper into the beaker and support it so that it stands upright but does not touch the sides. It is essential that the solvent does not wash away the coloured spots.

6. The solvent rises up the paper and separates out the pigments in the mixture.

7. You can see if any of the pigments from the mixture match the colour and heights of the pure pigments that you included in your experiment.

8. In this experiment, we can conclude that the purple pigment was made from the blue pigment on the right, and the second of the two possible red pigments, because the height of those spots matches the two spots from the purple pigment.

Progress Check

1. State the method you would use to separate magnesium powder from water.
2. State the method you would use to separate sodium chloride (common salt) from water.
3. State the method you would use to separate ethanol (boiling point 78°C) from water.

9.4 Testing for purity

It is often important to work with very pure substances. Here are some examples of why.
- When food additives are made, impurities might be poisonous.
- When medicines are made, impurities might be harmful for the patient.
- When useful materials are produced, impurities might affect their properties.

For example, copper is used to make electronic components. If a copper wire contains more than one atom of an impurity in a thousand atoms of copper, it will not conduct electricity well enough to be used in things like computers and mobile phones.

Melting point and boiling point

A pure sample of an element or compound has a definite melting point and a definite boiling point, under normal atmospheric pressure. A mixture (caused by the presence of impurities) is likely to melt or boil at a different temperature to the pure substance and it may melt or boil over a range of temperatures.

A pure sample of ice melts at 0°C and will then boil at 100°C. If any impurity is added to the water, it will melt at a *lower* temperature and boil at a *higher* temperature. Salt (sodium chloride) is an example of a substance that can be added to cause this change. This is very useful, because adding salt to icy roads in the winter causes the ice to melt and it also prevents refreezing.

Other methods

Impurities can also be detected in samples using sensitive tests. Chromatography can be used to detect impurities in many chemicals. Advanced techniques that chemists use include mass spectrometry and infra-red spectroscopy.

Progress Check

1. A sample of water was tested by heating it until it had completely evaporated. There was a small amount of white solid residue left behind. Explain whether the water was pure.
2. A second sample of water was heated until it began to boil. The temperature of the water was 100°C. What does this suggest?

Worked questions

a) Which of the following diagrams represent pure substances?

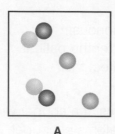 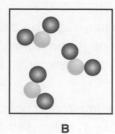 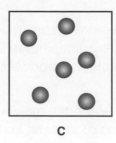 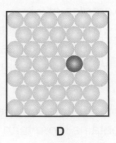

A   B   C   D

Write your answers here: *(2 marks)*

C and B

b) Which of the following are pure substances and which are mixtures? Put ticks in the correct boxes.   *(5 marks)*

	Pure?	Mixture?
Air		✓
Gold	✓	
Oxygen	✓	
Carbon dioxide	✓	✓
Orange squash	✓	

c) A chemical company uses hexane as a solvent in the preparation of new medicines. They need to make sure that their hexane is pure and doesn't contain any impurities. The scientists decide to test the boiling point of the hexane sample. Describe how they should do this test and how they will know if the hexane is pure or impure.   *(4 marks)*

They should heat the hexane until it boils and record the boiling point using a thermometer. Then they should compare their value with a data book or textbook. If their answer is the same, then their hexane is pure. If not, it isn't. A pure substance has a definite boiling point, but a mixture will boil over a range of temperatures.

Practice questions

1. This question is about sea water. Sea water is an example of a mixture. It is a solution of different salts dissolved in water. The most common salt in sea water is sodium chloride.

a) Match up the key word with its definition and the example in sea water.

Solute	The mixture of the solid and the liquid after dissolving	Sea water
Solvent	The solid that has dissolved	Water
Solution	The liquid	Salt

(4 marks)

b) Describe how you could extract pure, dry salt crystals from a mixture of sand and salt. You should describe each stage of the separation process with a labelled diagram and explain how it works and why you have included it in your plan. *(6 marks)*

c) In very hot countries, sea water is sometimes purified to make water for drinking. Drinking water must have no salt in it. Name a separation process that would produce pure water from drinking water. *(1 mark)*

d) Explain how this process works. Use a diagram in your answer. *(3 marks)*

2. This question is about testing for purity.

a) Explain why it is important to test for purity when making a medicine. *(1 mark)*

b) Medicines can be tested for purity using chromatography. Put the steps below in order, to describe how to make a chromatograph.

A Place the piece of paper into a beaker containing a small amount of solvent (e.g. water).

B A pure sample will be made from only one spot.

C Draw a pencil line across the bottom of the chromatography paper.

D Remove the paper and allow to dry.

E Put small spots of the substances to the tested onto the pencil line.

F Compare the height of the spots to identify any that match. *(5 marks)*

c) Distilled water should be pure. Explain how you can test a sample of distilled water to make sure it has no impurities. *(2 marks)*

Learning Summary

After completing this chapter you should be able to:

- write word and symbol equations to describe chemical reactions
- explain whether a reaction is exothermic or endothermic
- describe and identify combustion, oxidation and reduction reactions
- explain what a catalyst does.

10.1 Chemical equations

You can remind yourself about the difference between a chemical reaction and a physical change by looking back at Topic 8.3.

Chemical reactions are described using chemical equations.

Remember that a chemical equation always includes an arrow ($\rightarrow$) and never an equals sign (=) which is what you use when doing calculations in maths and in science. Use the plus (+) symbol instead of the word *and*. Physical changes cannot be represented by meaningful chemical equations, because no chemical change has taken place.

Writing word equations

In a word equation, the names of the chemicals are written out in words. For example, when magnesium metal reacts with dilute hydrochloric acid, magnesium chloride and hydrogen gas are produced. The word equation for this reaction is written like this…

Magnesium + hydrochloric acid $\rightarrow$ magnesium chloride + hydrogen

Note the following important points…

- The chemicals that react with each other are on the left of the arrow. These are called the **reactants**.
- On the right of the arrow are the **products**.
- Words like *metal, dilute* and *gas* are never included in word equations; only the names of the actual chemical elements and compounds.

Sometimes the information you need can be hard to extract from a written description of the reaction. Go through the information and highlight the reactants in one colour and the products in another colour. This will help. Ignore any descriptive words like *solid* or *hot*. For example:

Copper sulfate and water are produced in the reaction between warm sulfuric acid and powdered copper oxide.

Copper oxide + sulfuric acid $\rightarrow$ copper sulfate + water

Note that you can put copper oxide and sulfuric acid in either order, as long as they are on the left of the arrow. Likewise, you could swap copper sulfate with water.

Symbol equations

To save time and space, chemists often write the symbols and formulae for elements and compounds rather than their names. This means that chemical equations can be written with symbols instead of words. This is useful because it allows you to see which atoms have reacted.

Chlorine (Cl_2) reacts with methane (CH_4) to make chloromethane (CH_3Cl). The other product is hydrogen chloride gas (HCl). We can write a word equation and then a symbol equation underneath.

Chlorine + Methane ⟶ Chloromethane + Hydrogen chloride

Cl_2 + CH_4 ⟶ CH_3Cl + HCl

In another reaction, water (H_2O) is the only product in the reaction between hydrogen and oxygen. Here is the word and symbol equation.

Hydrogen + Oxygen ⟶ Water

H_2 + O_2 ⟶ H_2O

Sometimes, **state symbols** are included in symbol equations to show whether a chemical is a solid, liquid, gas or **dissolved** in water. We use s for solid; l for a pure liquid; g for a gas; and aq for aqueous, which means dissolved in water. For example:

Mg (s) + HCl (aq) → $MgCl_2$ (aq) + Cl_2 (g)

Balancing chemical equations

Looking at some of the particle diagrams above, we can see that, in some reactions, it is hard to see how atoms are conserved. Sometimes, it seems that atoms must be destroyed or created to make the products. **This never happens!** The problem is that we have not yet **balanced** the chemical equation.

It is important to learn how to balance symbol equations because otherwise they don't really make sense. Here is a simple but effective way to balance any chemical equation. A worked example is included alongside the steps.

Cut out some circles from coloured card and label them with the symbols for carbon, hydrogen and oxygen. Model both of the following reactions.

$CH_4 + O_2 \rightarrow CO_2 + H_2O$

$CO + H_2 \rightarrow CH_3OH$

Conservation of mass is explained in Topic 8.3.

1. Write out the symbol equation, draw a line underneath and a dotted line down from the arrow.

$$CH_4 + O_2 \rightarrow CO_2 + H_2O$$

2. Write out the formulae underneath so that you can clearly see how many of each type of atom there are in each compound or element. It is helpful if you do this in a new colour.

$$CH_4 + O_2 \rightarrow CO_2 + H_2O$$
$$CHHHH \quad OO \quad | \quad COO \quad HHO$$

> Make a model to help you to teach a parent or friend how to balance chemical equations. You could include the atoms and a see-saw to help them to understand.

3. Count how many C atoms there are under the line on the left and check that there is the same number on the right. In this example, the C atoms are already balanced. Repeat this with the next element, H. If you need more on the right, you must write them in on another line but the two new H atoms must come bonded to an O (you must write HHO). The hydrogens are now balanced.

$$CH_4 + O_2 \rightarrow CO_2 + H_2O$$
$$CHHHH \quad OO \quad | \quad COO \quad HHO$$
$$\qquad\qquad\qquad\qquad\qquad HHO$$

4. Now count up the number of O atoms. You will notice that there are two on the left and four on the right, so you need to add another two O atoms on the left. Next, check all the atoms again to make sure they are still balanced.

$$CH_4 + O_2 \rightarrow CO_2 + H_2O$$
$$CHHHH \quad OO \quad | \quad COO \quad HHO$$
$$\qquad\qquad OO \quad | \qquad\qquad HHO$$

5. Finally, count the number of lines you have used for each chemical (in red, under the line). This tells you what number to put in front of each chemical. You don't need to use the number 1.

$$CH_4 + 2O_2 \rightarrow CO_2 + 2H_2O$$
$$CHHHH \quad OO \quad | \quad COO \quad HHO$$
$$\qquad\qquad OO \quad | \qquad\qquad HHO$$

Note: you never change the small numbers in any formula whilst trying to balance an equation.

Progress Check

1. Hydrogen peroxide (H_2O_2) decomposes (breaks down) to produce water (H_2O) and oxygen (O_2). Write a word equation for this reaction.
2. Write a symbol equation for this reaction.
3. Balance the symbol equation for this reaction.

10.2 Energy in chemical reactions

We learned in Chapter 8 (section 8.3) that chemical reactions often release energy but they can take it in as well. If a reaction releases heat, energy is transferred from the chemical store to the thermal store of the surroundings.

> You can read more about different energy stores in Topic 15.1.

Exothermic reactions

Most chemical reactions release energy to the surroundings. These are called **exothermic** reactions. The energy lost by the chemicals is usually in the form of heat, so they feel hot to the touch. Some exothermic reactions give off light without any heat energy, like glow sticks.

Some examples of exothermic reactions include...

Burning any fuel

Reaction of magnesium with acid

A glow stick

Adding water to anhydrous copper sulfate

Endothermic reactions

Reactions that absorb energy from the surroundings are described as **endothermic**. The energy absorbed is usually in the form of heat, so these reactions feel cold. Sometimes, the energy is absorbed in the form of light, like in **photosynthesis**. Endothermic reactions are less common than exothermic reactions. Examples include eating sherbet or coffee biscuits, which feels cold on the tongue.

Photosynthesis

Measuring energy changes

You can usually tell if a reaction is exothermic or endothermic by using a thermometer. Take the temperature of the surroundings before the reaction and after the reaction. If the surroundings get hotter (or feel hotter), the reaction is exothermic. If the reaction is taking place in a solution, you can put the thermometer into the solution itself because the solvent (water) is part of the surroundings.

Combustion

When things burn, they release energy, so **combustion** (burning) is an example of an exothermic reaction.

A substance that is burned specifically to release useful heat energy is called a **fuel**. Fuels are examples of flammable substances and will usually carry the **flammable hazard symbol.**

For combustion to take place, three things are needed. We use the **fire triangle** to help us to remember this.

If one of the sides of the triangle is removed, combustion is not possible. For example, if you stop oxygen from reaching a fire, the fire will go out. This is how carbon dioxide fire extinguishers work. If you cool a fire down by spraying it with water, it will also go out.

Complete and incomplete combustion

When there is a good supply of oxygen, a fuel that contains carbon and hydrogen (like methane, CH_4) burns to form carbon dioxide and water and it releases a lot of energy. This is what happens when the air hole is open on a Bunsen burner and it is called **complete combustion**. When there is a poor supply of oxygen (such as when the air hole is closed on a Bunsen), **incomplete combustion** occurs, which releases less energy. This produces dirty soot (carbon) and poisonous carbon monoxide (CO).

Thermal decomposition

When some compounds are heated very strongly, they break down to form other compounds or sometimes elements. This type of reaction is called **thermal decomposition**. Because the chemicals take in energy in this reaction, it is an **endothermic** reaction.

You might have investigated thermal decomposition in your science lessons. A good example is copper carbonate, $CuCO_3$, which is a green powder. When heated strongly in a boiling tube, it turns black and it looks as if it is boiling, as the gas produced in the reaction pushes the powder particles around in the boiling tube. Look at the equation below and identify the gas that is produced in this reaction.

Copper carbonate → copper oxide + carbon dioxide

$$CuCO_3 \rightarrow CuO + CO_2$$

Mass changes in thermal decomposition reactions

Would you expect the mass of the chemical in the boiling tube before the reaction to be more or less than the mass of the chemical at the end of the reaction? Explain your answer, using ideas about atoms and conservation of mass. Refer to Topic 8.3 for help if you need to. Try to plan an investigation to find out how the mass changes when you start with different masses of copper carbonate.

Make a model of the fire triangle, with removable sides so that you can use it to explain three ways to stop a fire from burning.

1. A thermometer is placed into some water in a boiling tube. Some calcium metal is added and it starts to fizz. The temperature rises. Is the reaction exothermic or endothermic?
2. Respiration is the chemical process that takes place inside every living cell to provide the organism with energy for life processes. Is respiration exothermic or endothermic?
3. Calcium carbonate ($CaCO_3$) thermally decomposes when it is heated very strongly, in a similar reaction to copper carbonate. Suggest the names of the two compounds produced in this reaction.

Progress Check

10.3 Oxidation and reduction

There are several ways to define an **oxidation** reaction. The simplest is to say that something is oxidised if it gains oxygen in a reaction.

So, whenever you see the element oxygen as one of the reactants, you are looking at an oxidation reaction. For example:

Burning methane	Methane + oxygen → carbon dioxide + water $CH_4 + 2O_2 \rightarrow CO_2 + 2H_2O$	
Respiration	Glucose + oxygen → carbon dioxide + water $C_6H_{12}O_6 + 6O_2 \rightarrow 6CO_2 + 6H_2O$	
Burning magnesium	Magnesium + oxygen → magnesium oxide $2Mg + O_2 \rightarrow 2MgO$	

Reduction

The opposite of oxidation is **reduction**. This happens when a chemical loses oxygen. We say that it has been *reduced*. Oxidation and reduction often take place together in the same reaction, when one substance loses oxygen and another one gains it. This type of reaction is often called **redox**.

Oxidation can also be identified when something has gained oxygen from another compound. In each of these reactions, the element being oxidised is highlighted in red. The compound being reduced (the source of the oxygen) is highlighted in blue.

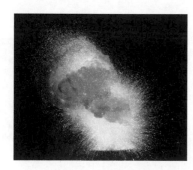

Thermite reaction	Iron oxide + aluminium → aluminium oxide + iron $Fe_2O_3 + 2Al \rightarrow Al_2O_3 + 2Fe$
Extracting copper	Copper oxide + carbon monoxide → copper + carbon dioxide $CuO + CO \rightarrow Cu + CO_2$

Progress Check

1. Iron reacts with oxygen in the air when rust is formed. Is this process oxidation or reduction?
2. When iron is extracted from iron ore, the reaction is $Fe_2O_3 + 3CO \rightarrow 2Fe + 3CO_2$. Which substance has been oxidised and which has been reduced?

10.4 The speed of reactions and catalysts

🎧 33

Some chemical reactions are very fast, like explosions and burning. Some reactions are very slow, like rusting. It is very important for chemists to be able to control the speed of chemical reactions. This is because…

- If the reaction is too fast it might be dangerous.
- If the reaction is too slow, you might not produce the chemical quickly enough to make enough money from selling it.

The speed of a chemical reaction depends on a number of factors. You can speed up a chemical reaction by doing one or more of the following…

- Increase the temperature of the reactants.
- Increase the concentration of any dissolved reactants.
- Increase the pressure of any gaseous reactants.
- Grind up any lumps of solid into powder.
- Add a catalyst.

A **catalyst** is a chemical that speeds up a reaction without being used up or chemically changed. Catalysts are very useful chemicals and are used in many chemical processes to save time and therefore money. They also allow a reaction to take place more effectively at lower temperatures, which also saves money.

In a small group, devise and perform a role play that explains what catalysts do to chemical reactions. Remember that catalysts do not increase the temperature, so the particles don't move any quicker, but catalysts do help the particles to collide and react more quickly.

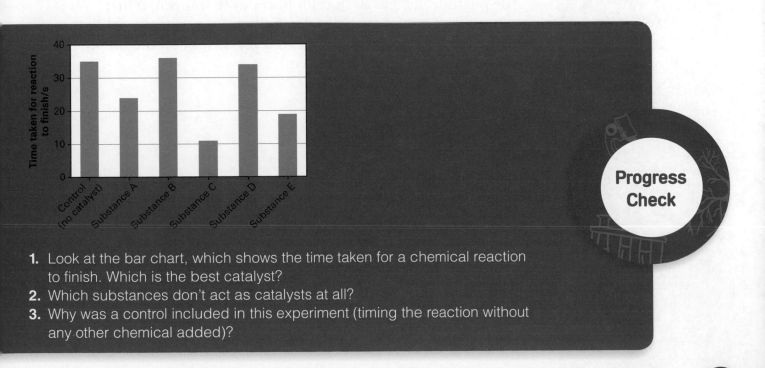

Progress Check

1. Look at the bar chart, which shows the time taken for a chemical reaction to finish. Which is the best catalyst?
2. Which substances don't act as catalysts at all?
3. Why was a control included in this experiment (timing the reaction without any other chemical added)?

Worked questions

a) For this type of question, there is often one mark for the reactants and one mark for the products. Remember to ignore words like 'dilute', 'hot', 'solid', because this might lose you a mark.

Remember to use an arrow ($\rightarrow$) in a chemical equation, not an equals sign ($=$). This will definitely lose you a mark.

b) Correct. The reactants are on the left of the arrow in a correct equation. Hydrochloric acid would also have earned the mark.

c) This question is testing your ability to recognise state symbols. (s) means solid, (l) means liquid, (g) means gas and (aq) means aqueous. The clue is in the production of CO_2 which is a gas, so you would see fizzing/bubbles. One mark for the predicted observation and one mark for referring to the correct product from the equation.

d) Balancing equations can be tricky but it is essential that you remember to never change the little numbers after a formula. Balancing equations ends with you writing big numbers **in front** of the formulae. In this case, you need a big 2 in front of the HCl.

e) Three marks are awarded here because the student has correctly stated the three things needed for fire. The final mark is awarded for using scientific ideas to explain how fires can be put out.

a) Sodium chloride is a salt. It can be made by reacting dilute sodium hydroxide (an alkali) with dilute hydrochloric acid. Water is also produced in the reaction. Write a word equation for this reaction. *(2 marks)*

Sodium hydroxide + hydrochloric acid → sodium chloride + water

b) Write down the name of one of the reactants. *(1 mark)*

Sodium hydroxide

c) Sodium chloride can also be made in the reaction between sodium carbonate (Na_2CO_3) and hydrochloric acid (HCl). Look at the symbol equation below.

$$Na_2CO_3 \text{ (s)} + 2HCl \text{ (aq)} \rightarrow 2NaCl \text{ (aq)} + H_2O \text{ (l)} + CO_2 \text{ (g)}$$

Describe what you would expect to **see** when this reaction is taking place. Refer to the state symbols in your answer. *(2 marks)*

Bubbles of gas, which is the carbon dioxide.

d) Hydrochloric acid also reacts with magnesium, to form magnesium chloride and hydrogen. Here is the symbol equation for this reaction.

$$HCl + Mg \rightarrow MgCl_2 + H_2$$

Balance the symbol equation. Write your answer on the line below. *(2 marks)*

$2HCl + Mg \rightarrow MgCl_2 + H_2$

e) Combustion is another name for burning. Three things are needed for burning and these are represented in the fire triangle. State the three things that are needed and explain why this is useful information for fire fighters. *(4 marks)*

The fire triangle tells us that for burning to happen, a fuel, oxygen and heat are needed. Firefighters find this information useful because they can put fires out by cutting off the supply of oxygen or by cooling them down.

Practice questions

1. This question is about limestone.

    Limestone is a very important raw material in the chemical industry. Chemicals used from limestone are used in toothpaste, food, cosmetics, farming and building. Limestone is mostly made from the compound calcium carbonate, $CaCO_3$.

    a) When heated, calcium carbonate ($CaCO_3$) breaks down to form calcium oxide (CaO) and carbon dioxide. Write a word equation for this reaction.   *(2 marks)*

    b) Write a symbol equation for this reaction.   *(2 marks)*

    c) What type of chemical reaction is this? Choose from the list:

    - Electrolysis

    - Oxidation

    - Thermal decomposition

    - Displacement   *(1 mark)*

    d) One of the products of this reaction, calcium oxide, reacts violently when small amounts of water are added, releasing so much heat that the water often boils off as steam. What term is used to describe reactions that release heat into the surroundings?   *(1 mark)*

    e) Calcium chloride is produced when calcium carbonate reacts with hydrochloric acid. The other products are water and carbon dioxide. Write a word equation for this reaction.   *(2 marks)*

    f) The carbon dioxide produced in this reaction can be detected using a simple test. Describe how to do this test and what the positive result is.   *(2 marks)*

    g) A student in your class suggests that this reaction might be affected by a catalyst. Describe what a catalyst is.   *(2 marks)*

    h) Describe an experiment to allow you to test the student's idea. In your plan, state the independent variable, the dependent variable and how you will make it a valid (fair) test.   *(4 marks)*

2. This question is about burning fuels.

   a) Combustion (burning) is an example of an oxidation reaction. State what is meant by an oxidation reaction.

   (1 mark)

   b) A student investigated a number of fuels by burning them and using the energy released to heat up some water in a metal beaker, which was clamped above the burning fuel. List four factors that the student should have controlled to make this a fair test.

   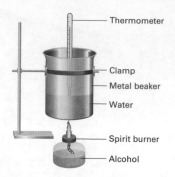

   Thermometer
   Clamp
   Metal beaker
   Water
   Spirit burner
   Alcohol

   (4 marks)

   c) The student measured the temperature rise of the water in the beaker for each fuel. Here are the results. Calculate the temperature change for propanol.

Fuel	Temperature of water before heating (°C)	Temperature of water after heating (°C)	Temperature change (°C)
Methanol	19	34	15
Ethanol	20	37	17
Propanol	19	38	
Butanol	19	45	26

   (1 mark)

   d) Plot a suitable graph to represent the data in the table.

   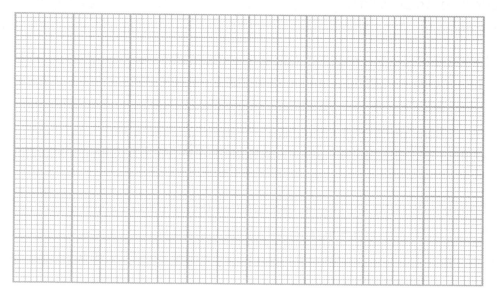

   (5 marks)

   e) Which is the best fuel? Explain your answer.

   (2 marks)

   f) Another student from the same class suggested that a lid should be put on to the top of the beaker. Explain why.

   (1 mark)

After completing this chapter you should be able to:
- describe the pH scale and how it is used to classify acids, alkalis and neutral substances
- use word equations to predict the products of the reactions of acids with alkalis and other compounds
- describe how to produce and extract a solid sample of a particular salt.

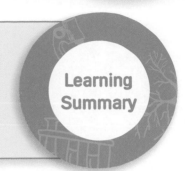

Learning Summary

11.1 Acids, alkalis and the pH scale 34

The pH scale is used to measure whether a solution is acidic, neutral or alkaline. The most commonly used section of the pH scale goes from 0 to 14, as seen in the diagram below. The lower the pH of a solution, the more acidic it is. Solutions with a pH of 7 are neutral. A pH above 7 means that a solution is alkaline. Acids always contain hydrogen. Alkalis are usually soluble metal hydroxides.

pH can be measured using a pH probe, which is sometimes connected to a computer. It can also be approximately measured using a full-range indicator, like **universal indicator**. When using an indicator, you must compare the colour of the indicator with the colour chart for that indicator. The typical colours of Universal Indicator are shown in the diagram below.

Scientists often try to classify things that they are studying. This involves putting things into groups that share characteristics, so they have something in common. For example, mammals have many shared characteristics, so it makes sense to study them as one group. Classifying solutions as **acids**, **alkalis** or **neutral** solutions allows us to **make predictions** about how they will react.

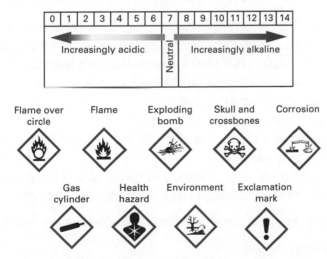

Other indicators can tell you whether something is acidic or alkaline but not usually how acidic or alkaline a solution is. For example, litmus is red in acids and blue in alkalis; phenolphthalein is colourless in acids and neutral solutions but bright pink in alkalis.

1. Our stomach contents have a pH of around 1 or 2. Is this acidic or alkaline?
2. Ethanol solution is neutral. What is its pH?
3. Ammonia solution is a weak alkali. Suggest its pH.

Progress Check

11.2 Reactions of acids

Acids and alkalis can both be dangerous if they are concentrated and if proper safety precautions are not followed when using them. However, many everyday substances are slightly acidic or alkaline and we don't worry too much about them! For example, vinegar is an acid with a pH around 4 and sea water is slightly alkaline.

Safety precautions in the lab

When they are very **concentrated**, **both acids and alkalis** can be corrosive. This means that they can attack and 'eat through' other substances, including skin. As well as the normal safety precautions of wearing goggles during every practical, chemical-resistant gloves should be worn and sometimes a face shield will be used instead of goggles.	 Corrosive hazard symbol
When they are more **dilute**, **both acids and alkalis** are likely to be irritants. Irritants may make your skin sore but they are less dangerous than corrosive substances. Gloves are not normally worn but splashes on your skin should be washed off.	 General hazard warning, as found on irritants

Reactions of acids with alkalis

We can consider acids and alkalis to be chemically opposite, so they react together to form a neutral solution, as long as the right quantities are used. This type of reaction is called a **neutralisation** reaction. Here is the procedure for one way of doing a neutralisation reaction.

1. Place some acid solution into a beaker.
2. Add two drops of universal indicator. It will turn red.
3. Slowly add some alkali to the beaker, stirring with a glass rod. The colour should change to green.
4. If the indicator turns blue or purple, you have added too much alkali. Stop, and add some acid to get back to a neutral solution.

The general word equation for an acid–alkali neutralisation reaction is…

acid + alkali → a salt + water

A salt is a chemical that is made when an acid is neutralised. The most common salt is sodium chloride but there are many others.

For example:

hydrochloric acid + potassium hydroxide → potassium chloride + water

sulfuric acid + sodium hydroxide → sodium sulfate + water

nitric acid + lithium hydroxide → lithium nitrate + water

Make a colourful poster of the pH scale and stick pictures of common household substances on to it in the appropriate places, showing their pH values.

Reactions of acids with metals

Dilute acids will react with metals that are fairly reactive. Unreactive metals like copper, silver, gold and platinum will not react with most dilute acids.

You can find out more about how reactive different metals are in section 13.1.

When a metal reacts with an acid, the products are a salt and hydrogen gas. You see bubbles of hydrogen being produced in the reaction mixture at the bottom of the boiling tube. You can trap this gas using a cork/bung or with your thumb (wearing protective gloves) and test it with a burning splint.

It is useful to be able to identify gases produced in reactions. Hydrogen gives a squeaky pop with a burning splint. Carbon dioxide turns limewater cloudy. It is unusual for oxygen to be produced in chemical reactions but it can be detected because it relights a glowing splint.

The general equation for the reaction between a metal and an acid is…

acid + metal → a salt + hydrogen

For example:

hydrochloric acid + zinc → zinc chloride + hydrogen

sulfuric acid + magnesium → magnesium sulfate + hydrogen

nitric acid + calcium → calcium nitrate + hydrogen

How to name salts

There are patterns in naming salts, so you can work out exactly what salt will be produced in a given reaction. Look at the word equations on this page and the previous page. The first part of the name of the salt always comes from the alkali or the metal. The second part comes from the acid…

hydrochloric acid always makes salts that end in **chloride**

sulfuric acid always makes salts that end in **sulfate**

nitric acid always makes salts that end in **nitrate**

For example:

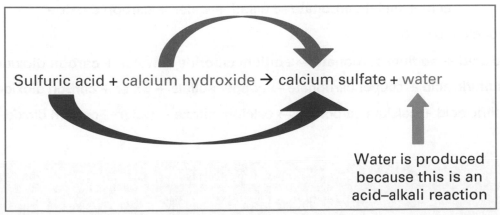

Sulfuric acid + calcium hydroxide → calcium sulfate + water

Water is produced because this is an acid–alkali reaction

Reactions of acids with metal oxides

Acids can be neutralised by metal oxides as well. Most metal oxides are insoluble in water. The products of these reactions are the same as for acid–alkali neutralisations: a salt and water.

Reactivity of different metals with hydrochloric acid; calcium and magnesium vigorous, zinc less so, copper no reaction

The general equation is...

acid + metal oxide → a salt + water

For example:

hydrochloric acid + copper oxide → copper chloride + water

sulfuric acid + magnesium oxide → magnesium sulfate + water

nitric acid + zinc oxide → zinc nitrate + water

Reactions of acids with metal carbonates

Acids can be neutralised by metal carbonates as well. Most metal carbonates are insoluble in water but lithium, sodium and potassium carbonates do dissolve in water. The products of these reactions are a salt, water and carbon dioxide gas.

The general equation is...

acid + metal carbonate → a salt + water + carbon dioxide

For example:

hydrochloric acid + sodium carbonate → sodium chloride + water + carbon dioxide

sulfuric acid + copper carbonate → copper sulfate + water + carbon dioxide

nitric acid + calcium carbonate → calcium nitrate + water + carbon dioxide

Produce a card sorting activity that allows you to test your ability to work out the reactants and products for a variety of acid reactions. Remember to include some + and → symbols so that you can make full word equations.

Progress Check

1. Write a word equation for the reaction of nitric acid with lithium hydroxide.
2. Write a word equation for the reaction of hydrochloric acid with aluminium.
3. Write a word equation for the reaction of sulfuric acid with iron oxide.
4. Write a word equation for the reaction of nitric acid with nickel carbonate.

11.3 Producing a salt

36

In Topic 11.2 we saw that there are several different methods for producing a salt: acid + alkali, acid + metal; acid + metal oxide; and acid + metal carbonate. To make a given salt, you need to choose the right acid and also a suitable metal, alkali, metal oxide or metal carbonate.

For example, to make sodium **nitrate**, you need to use **nitric** acid. The sodium part of the salt could come from sodium hydroxide (an alkali), or sodium carbonate. Using sodium metal would be too dangerous and sodium oxide is not a stable compound.

Practical method

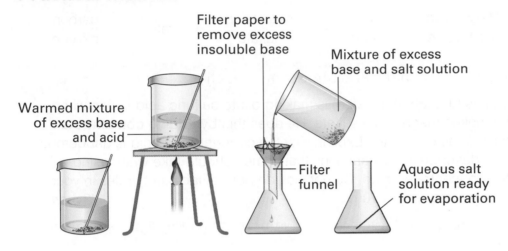

Filter paper to remove excess insoluble base

Mixture of excess base and salt solution

Warmed mixture of excess base and acid

Filter funnel

Aqueous salt solution ready for evaporation

> You will learn more about bases in Topic 12.2.

1. Take a measured amount of your dilute acid, e.g. 20 cm³.

2. Add small amounts of metal, alkali, metal oxide or metal carbonate. If you are using a metal or metal oxide, you may need to warm the acid gently. If you are using a metal or a carbonate, you should see bubbles.

3. Use a glass rod to stir the mixture and then transfer a drop of the solution on to a piece of universal indicator paper. If the solution is still acidic, you need to add more metal, metal oxide, alkali or carbonate until it is neutral.

4. If there is any solid still remaining in the mixture, filter it out and use the **filtrate** for the next step.

5. Evaporate the water from the mixture to give crystals of the pure salt. To do this, pour the salt solution into an evaporating basin on a gauze on a tripod. Heat gently until half the water has evaporated. Leave the basin in a warm place to evaporate the rest.

> Without looking at your notes or this page, try to mime the procedure for making a salt from an acid and a metal oxide. Then check to see if you included all the steps.

1. Suggest two compounds that you could react together to make iron sulfate.
2. Why are bubbles seen when an acid reacts with a metal or metal carbonate but not with an alkali or metal oxide?

Progress Check

Worked questions

a) One mark is awarded for correctly stating that hydrogen is produced when a metal reacts with a dilute acid. The second mark is for the rest of the equation being correct.

b) The student is right; you would see bubbles during this reaction. The second mark is for explaining that the bubbles are carbon dioxide gas (as seen from the equation).

c) The first mark is for saying that concentrated acids are corrosive. Safety glasses are inadequate when using corrosive liquids. Goggles are better but a face shield should really be worn.

d) A simple question like this requires a short answer. There is no need to repeat the question. Usually, a range of answers will be accepted for a question that asks you to suggest something. Here, any number below 4 would probably be fine.

e) A perfect answer! One mark is awarded for the reactants and one for the products.

a) Magnesium sulfate is sometimes known as 'Epsom salts'. There are several ways to make magnesium sulfate. In one method, magnesium metal is reacted with dilute sulfuric acid. A gas is given off in the reaction. Write a word equation for this reaction. *(2 marks)*

Magnesium + sulfuric acid → magnesium sulfate + hydrogen

b) Magnesium sulfate can also be made from magnesium carbonate. Look at the word equation below and describe what you would **see** during this reaction. Explain your answer. *(2 marks)*

Magnesium carbonate + sulfuric acid → magnesium sulfate + water + carbon dioxide

You would see bubbles because of the carbon dioxide.

c) James (the lab technician) is making dilute sulfuric acid to use when he makes magnesium sulfate. He does this by adding **concentrated** sulfuric acid to water. Explain why the concentrated acid is dangerous and describe the precautions that James should take when using the concentrated acid. Do not include normal lab safety rules in your answer (e.g. do not run). *(3 marks)*

The concentrated acid is corrosive. James should wear a face shield and gloves.

d) Suggest a pH for concentrated sulfuric acid. *(1 mark)*

1

e) Magnesium sulfate can also be made using magnesium hydroxide. Write a word equation for this reaction. *(2 marks)*

Magnesium hydroxide + sulfuric acid → magnesium sulfate + water

Practice questions

1. This question is about making copper sulfate.

a) Copper sulfate is a salt that can be made from copper oxide and an acid.
   Name the acid. *(1 mark)*

b) Suggest a pH for the acid you have chosen and also for copper sulfate solution. *(2 marks)*

c) Copper oxide is a black powder that is insoluble. Describe the steps that you would follow to produce a solution of copper sulfate. Include in your answer the apparatus you would use and how you would ensure that the copper sulfate solution is neutral. *(4 marks)*

d) Having prepared your copper sulfate solution, describe how you would use this to make some solid dry crystals of copper sulfate. Use a labelled diagram in your answer. *(4 marks)*

e) Copper sulfate is **harmful** and **dangerous for the environment**. Other copper compounds are toxic. Which is the correct hazard symbol to show that something is toxic? *(1 mark)*

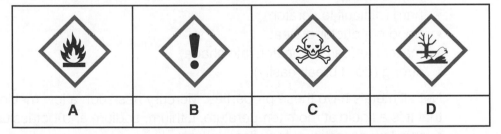

2. This question is about the pH scale.

a) On the diagram below of the pH scale, add the words: **neutral, alkali, acid**. *(3 marks)*

0	1	2	3	4	5	6	7	8	9	10	11	12	13	14	15

b) Universal indicator can be used to test the pH of a solution. State the colour of universal indicator in the following solutions. *(3 marks)*

pH 1   pH 7   pH 14

After completing this chapter you should be able to:
- describe the typical properties of metals and non-metals and use them to explain uses of different substances
- explain how scientists used ideas and evidence to develop the periodic table
- describe the structure of the periodic table and use the patterns in it to make predictions about the properties of elements.

12.1 Properties and uses

A **property** is a characteristic of a substance. For example, *high melting point* is a property of some substances. Another example of a property is *good conductor of electricity*. Remember that *cheap* is not a property, because it depends on supply and demand in the global market.

Metals

Here are the typical properties of most metals…
- high melting point
- good **conductor** of electricity
- shiny
- ductile (can be drawn into wires)
- high **density**
- hard (difficult to scratch)
- good conductor of heat
- malleable (can be beaten into shape)
- strong (don't break easily).

Not all metals have these properties. Mercury has such a low melting point that it is a liquid at room temperature. Lithium, sodium and potassium have such a low density that they float on water.

Metals are sometimes mixed together to form an **alloy**. Alloys have properties that are different from the elements they are made from. Steel is an alloy of iron with carbon and sometimes other elements.

The properties of metals help to explain their uses. Here are some examples.

	Metal	Used for...	Relevant properties...
	Copper	Electrical wiring in the home	Very good conductor of electricity; ductile
	Steel	Building bridges and skyscrapers	Strong; malleable
	Aluminium	Saucepans	Good conductor of heat; high melting point
	Brass (an alloy of copper and zinc)	Door handles, bells	Shiny

When metals react with oxygen, they form metal oxides. These compounds are nearly always **bases**, which means that they are chemicals that can neutralise acids. If a base dissolves in water then it is called an alkali.

Non-metals

Non-metal elements are often gases; for example, hydrogen and oxygen. This is because these elements have a very low boiling point. Other non-metal elements are solids with a very high melting point; for example, carbon and silicon. The reason for this big difference in melting and boiling points can be explained by the bonding and arrangement of the atoms in these elements. On the next page is a summary of the properties of some non-metal elements...

You can find out more about the reactions of acids with metal oxides and alkalis in section 8.2. You will also be able to see some word equations for the reactions of acids with metal oxides, to form a salt plus water.

Property	Notes
Very low boiling point (therefore a gas)	Examples include: hydrogen, helium, nitrogen, oxygen, fluorine, neon, chlorine, argon
Low melting point (therefore a solid that melts quite easily)	Examples include: phosphorus, sulfur
Very high melting point (therefore a solid that doesn't melt easily)	Examples include: boron, carbon (graphite, diamond), silicon
Usually very poor conductors of electricity	Some non-metals are semiconductors (e.g. silicon, used in electronic devices). Graphite (a form of pure carbon) is the only non-metal element that is a very good conductor of electricity.
Brittle	If a non-metal element is solid, it will be brittle, which means that it will shatter when hit with a hammer. This is the opposite of metals.

Non-metal elements are usually used in compounds rather than as pure elements. This changes their properties, as we saw in Topic 8.2. However, here are some non-metal elements that are useful because of their properties.

	Element	Used for…	Relevant properties
	Hydrogen	As a clean fuel because it produces only water when it is burned. The picture shows a hydrogen-powered racing car	Highly flammable
	Neon	Neon lighting	Gives out light energy when an electrical current passes through it
	Helium	Party balloons	Very low density

The oxides of non-metals are usually acidic. For example, carbon dioxide dissolves in water to form carbonic acid. Sulfur oxides dissolve in rain water to form acid rain. Nitrogen oxides dissolve in water to form nitric acid.

Write out all the property words you can think of on separate cards. Then jumble them up and try to sort them into two groups: metals and non-metals. Do any cards belong in both piles?

Progress Check

1. Suggest why aluminium is used for high-voltage mains electricity cables.
2. Iodine is a solid with a low melting point. It does not conduct electricity. Is it a metal or non-metal?
3. Hydrogen and helium are both gases with a very low density. Suggest why helium is used for party balloons instead of hydrogen.

12.2 Developing and using the periodic table

Who developed the periodic table?

Several scientists helped to develop the **periodic table**. They did this by looking for patterns in the way that elements reacted and by using measurements of atomic mass. Scientists tell each other about their ideas so that they can be reviewed and tested.

Döbereiner was a German chemist who, in 1829, noticed that some elements had very similar chemical properties. He put some elements into groups of three, which he called **triads**. For example, lithium, sodium and potassium are all soft metals that float on water and react quickly with it to produce hydrogen and an alkaline solution.

Johann Wolfgang Döbereiner

Newlands was an English chemist who proposed the **Law of Octaves** in 1865. He had arranged the elements in order of increasing atomic mass and noticed that there was a repeating pattern of properties. This reminded him of the scales used in music, hence the term *octaves*. Other scientists did not like this comparison and Newlands was ridiculed by them.

John Newlands

Mendeleev was a Russian chemist who also noticed the repeating properties of elements when they were arranged in order of increasing atomic mass. However, Mendeleev's table was much more popular and successful than previous versions because…

Dmitri Mendeleev

- Mendeleev recognised that not all the elements had been discovered, so he left gaps in his table so that other elements would fit into a logical pattern of properties.
- Mendeleev predicted the properties of these undiscovered elements. These predictions were later proved to be very accurate.

Scientists usually build their ideas on the work of other scientists. However, it appears that Dmitri Mendeleev came up with his ideas about repeating patterns in the properties of elements all by himself. Another scientist called Lothar Meyer came up with a very similar periodic table just after Mendeleev. The credit for the modern periodic table is generally given to Mendeleev because he got there first and because he used his table to predict the properties of undiscovered elements.

Monument to Mendeleev in St Petersburg, Russia, showing his original periodic table

The structure of the periodic table

Here is the modern periodic table, which shows the elements arranged in order of increasing atomic number. You will find a larger copy on the inside back cover. The zigzag line separates the metals (to the left) from the non-metals (to the right of the line).

Group 1	Group 2					Transition metals						Group 3	Group 4	Group 5	Group 6	Group 7	Group 0
							1 H 1										4 He 2
7 Li 3	9 Be 4				Period 2							11 B 5	12 C 6	14 N 7	16 O 8	19 F 9	20 Ne 10
23 Na 11	24 Mg 12											27 Al 13	28 Si 14	31 P 15	32 S 16	35.5 Cl 17	40 Ar 18
39 K 19	40 Ca 20	45 Sc 21	48 Ti 22	51 V 23	52 Cr 24	55 Mn 25	56 Fe 26	59 Co 27	59 Ni 28	63.5 Cu 29	65 Zn 30	70 Ga 31	73 Ge 32	75 As 33	79 Se 34	80 Br 35	84 Kr 36
85 Rb 37	88 Sr 38	89 Y 39	91 Zr 40	93 Nb 41	96 Mo 42	98 Tc 43	101 Ru 44	103 Rh 45	106 Pd 46	108 Ag 47	115 Cd 48	119 In 49	122 Sn 50	128 Sb 51	128 Te 52	127 I 53	131 Xe 54
188 Cs 55	137 Ba 56	139 La 57	178 Hf 72	181 Ta 73	184 W 74	186 Re 75	190 Os 76	192 Ir 77	195 Pt 78	197 Au 79	201 Hg 80	204 Tl 81	207 Pb 82	209 Bi 83	210 Po 84	210 At 85	222 Rn 56
223 Fr 87	226 Ra 88	227 Ac 89															

The vertical columns are called **groups** and are numbered from 1 to 7, then group 0 on the far right-hand side. Group 2 is highlighted with a vertical blue arrow, as are groups 5 and 0. Some groups have their own name. For example, the elements in group 1 are called the **alkali metals**, the elements in group 7 are called the **halogens** and the elements in group 0 are called the **noble gases**.

The **transition metals** are shaded in green and are found between groups 2 and 3. Here you will find familiar metals like iron and copper, as well as the most dense and least reactive metals.

The horizontal rows are called **periods**. Period 2 has been highlighted for you. Be careful to remember that period 1 has only two elements in it: hydrogen and helium. Every element has a unique address in the periodic table. For example, sulfur is in period 3 and group 6.

Did you know? There are approximately five times as many metal elements as there are non-metals. However, two non-metals (hydrogen and helium) make up more than 99% of the atoms in the universe!

Make a set of cards that includes the names of the scientists who helped to develop the periodic table, as well as the things that they suggested. Place them all face down and turn two over. Do they match? If not, turn them back over and try again.

Using the periodic table to make predictions

Because we can study patterns in the properties of elements in the periodic table, we can make predictions about other elements based on where they are.

Li	Lithium	Quite soft, floats on water, reacts to form hydrogen and an alkaline solution
Na	Sodium	Soft, floats on water, reacts quickly to form hydrogen and an alkaline solution
K	Potassium	Very soft, floats on water, reacts very quickly to provide an alkaline solution and hydrogen gas, which burns spontaneously with a purple flame
Rb	Rubidium	???

You can see that all the elements in group 1 react with water to form hydrogen and an alkaline solution, so we would predict that rubidium would also float on water and react to produce similar products. We can also see that, as you go down the group, the metals are getting softer and more reactive, so we would predict that rubidium would be softer than potassium and even more reactive.

Predictions can be made going across periods as well as down groups but this is less common. The reason we can do this is because patterns in one period are usually repeated in the period below. Here are the elements in periods 2 and 3.

Group 1	Group 2	Group 3	Group 4	Group 5	Group 6	Group 7	Group 0
Lithium	Beryllium	Boron	Carbon	Nitrogen	Oxygen	Fluorine	Neon
Sodium	Magnesium	Aluminium	Silicon	Phosphorus	Sulfur	Chlorine	Argon

Here is a graph of the melting points of the elements in period 2. You can see that carbon has the highest melting point.

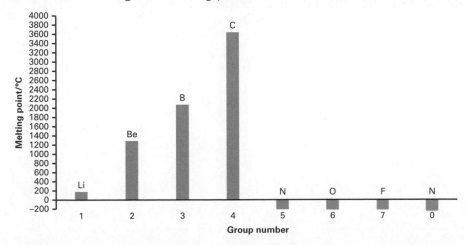

We would therefore predict that the same pattern of melting points is present in period 3, so we would expect silicon (below carbon, in group 4) to have the highest melting point. It does!

1. What element is in period 4 and group 7?
2. What group is aluminium in?
3. How would you expect caesium to react with water?

Progress Check

Worked questions

a) It is important to say whether you mean that something is a **good conductor of heat** or a **good conductor of electricity**. Without specifying which you mean, you won't get the marks. Rather than saying that a metal is shiny, it is usually better to explain **why** it is shiny. This will be because it is unreactive.

b) The definition given of an alloy is correct. Steel is a rather unusual alloy because it is a mixture of iron and carbon (a non-metal) but the answer is still correct.

c) Two marks are usually available for a word equation: one for the reactants and one for the products. Sometimes students write **air** instead of **oxygen** (which would not gain the mark).

d) An excellent answer. Knowing that metal oxides are bases allowed the student to deduce that they will react with acids. Another way to answer this would be to recall the use of copper oxide in making salts, when it reacts with acids.

e) The student has remembered that groups go vertically down and periods go horizontally. It is really important to remember that the first period contains only hydrogen and helium. Students often ignore this period when trying to find elements.

f) This answer would score the mark but there is no need to repeat the question in your answer.

a) Copper is used for a variety of purposes. Use ideas about properties to explain why copper is used in the following situations. *(3 marks)*

As the base for a saucepan: *it is a good conductor of heat.*

For electrical wiring: *it is a good conductor of electricity.*

Jewellery: *it is unreactive and therefore shiny.*

b) Bronze is an alloy of copper and tin. Explain what is meant by the term alloy. Give another example of an alloy in your answer. *(2 marks)*

An alloy is a mixture of two metals. Steel is an alloy.

c) Copper reacts with oxygen when it is heated in air. Write a word equation for this reaction. *(2 marks)*

Copper + oxygen → copper oxide

d) Suggest whether copper oxide will react with acids or alkalis. Explain your answer. *(2 marks)*

Copper oxide will react with acids because metal oxides are bases.

e) Use a periodic table to find the names of the following elements. *(3 marks)*

The element with the symbol Na: *sodium*

The element in group 2 and period 3: *magnesium*

The halogen in period 2: *fluorine*

f) What name is given to the elements in group 0 (sometimes called group 8)? *(1 mark)*

The elements in group 0 are called the noble gases.

Practice questions

1. This question is about elements in group 2 of the periodic table, which are called the **alkaline earth metals**. Mel reacted small pieces of each metal with dilute hydrochloric acid. Her results are shown in the table below.

   a)

Element	Temperature of acid at start	Temperature of acid after reaction	Observations
Beryllium	21	27	Small bubbles of gas are given off slowly
Magnesium	20	36	Bubbles of gas are given off quickly
Calcium	21	48	Rapid fizzing
Strontium			
Barium			

   What is missing from the boxes outlined with the thick black line? *(1 mark)*

   b) Suggest the results that Mel will get if she reacts strontium with dilute hydrochloric acid. *(2 marks)*

   c) The word equation for magnesium reacting with dilute hydrochloric acid is:

   Magnesium + hydrochloric acid → magnesium chloride + hydrogen

   Write a word equation for the reaction of barium with hydrochloric acid. *(2 marks)*

   d) What is the test for hydrogen gas? Describe what to do and what the positive result is. *(2 marks)*

2. This question is about how scientists developed the periodic table.

   a) Many scientists were involved in developing the periodic table. They all tried to spot patterns in the properties and reactions of the elements. Describe the contributions made by Döbereiner, Newlands and Mendeleev to the development of the periodic table. *(6 marks)*

   b) Scientists communicate with each other to share their ideas and to get credit for their work. How do scientists communicate with each other these days? *(2 marks)*

13.1 Constructing a reactivity series

We can use observations of the ways in which metals react with other substances to arrange them in order of reactivity. We can choose any metals to place into a reactivity series. For example: magnesium, calcium, zinc, potassium, copper, lithium and sodium.

Reactions with water

If small pieces of all of these metals are placed into separate beakers of water, the following observations are made.

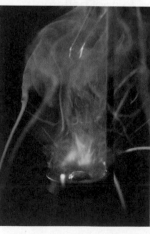

The reaction of potassium with water

Metal	Reaction with water
Magnesium	After several minutes, tiny bubbles appear on the surface of the metal.
Calcium	After a few seconds, the metal begins to fizz with tiny bubbles.
Potassium	Extremely fast reaction. The metal gives off sparks and a gas which burns.
Lithium	Fast reaction. The metal gives off a gas quickly.
Sodium	Very fast reaction. A gas is given off very quickly.
Zinc	No visible reaction.
Copper	No visible reaction.

Increasing reactivity ↑	
Potassium	We can use these observations to produce a reactivity series, as shown on the left. Note that, at this stage, we cannot put the bottom two metals into order of reactivity.
Sodium	
Lithium	
Calcium	
Magnesium	
Zinc and copper	

Reactions with dilute acids

To investigate the reactivity of the less reactive metals, we can react them with dilute acid. Here are the observations.

Metal	Reaction with dilute acid
	Magnesium Bubbles of gas produced very quickly. Reaction mixture feels hot.
	Copper No visible reaction.
	Zinc Tiny bubbles appear on the metal. Slow reaction.

This tells us that copper must be below zinc in the reactivity series and it confirms that magnesium is above zinc.

1. Platinum does not react with water, dilute acids or oxygen, even when heated. Where would you put it in a reactivity series?
2. Suggest why sodium was not added to dilute acid in the experiments described above.

Progress Check

40 **13.2** # Displacement reactions

In a displacement reaction, a more reactive element displaces (pushes out) a less reactive element from a compound. We can study **displacement reactions** of metals to test our understanding of the reactivity series. For example, we know that zinc is more reactive than copper, so when a piece of zinc is added to a solution of copper sulfate, the following reaction occurs.

Zinc + copper sulfate → zinc sulfate + copper

The zinc has displaced the copper, so metallic copper is formed in this reaction. The blue colour of the copper sulfate solution fades because zinc sulfate is colourless.

Here is a table of data about displacement reactions. See if you can work out which metal is the most reactive and which is the least reactive.

		Solution of metal compound		
		Copper sulfate	*Zinc sulfate*	*Magnesium sulfate*
Metal added	Copper		No reaction	No reaction
	Zinc	Copper displaced		No reaction
	Magnesium	Copper displaced	Zinc displaced	

We can see from the top row that copper cannot displace either zinc or magnesium, so it must be less reactive than both those metals. The bottom row tells us that magnesium displaces both copper and zinc, so magnesium is the most reactive of the three metals we tested.

It does not matter what the other part is in the metal compound: sulfate, nitrate, chloride, etc. So magnesium will displace zinc from zinc nitrate the same as it would from zinc sulfate.

A copper wire placed into silver nitrate solution. Over time, crystals of metallic silver appear on the wire, and the solution turns blue as it changes from silver nitrate to copper nitrate

> Write out the names of the metals featured so far in this chapter on to separate pieces of card. Arrange them into order of reactivity and then check to see if you got them all right. Devise a mnemonic (e.g. a rhyme or story) to help you to remember the order.

Progress Check

1. Use the reactivity series from section 13.1 to suggest a metal that would displace sodium from sodium chloride.
2. What would happen if copper metal was placed into a solution of calcium chloride?

13.3 Extracting metals

Increasing reactivity		
↑	Potassium	The majority of useful metals are found in the Earth's crust in compounds with other elements. For example, iron is usually bonded with oxygen in the compound iron oxide. We can extract the iron by displacing it with a more reactive element.
	Sodium	
	Lithium	
	Calcium	
	Magnesium	
	Aluminium	Look at the reactivity series, which includes two non-metals highlighted in red.
	Carbon	
	Zinc	
	Iron	Carbon can be used to displace any metal that is below carbon in the reactivity series. For example:
	Lead	
	Hydrogen	Carbon + iron oxide → iron + carbon dioxide
	Copper	
	Silver	Hydrogen can be used to displace copper from copper oxide but cannot be used to extract lead, iron or zinc.
	Gold	
	Platinum	

So the method of **extraction** for a metal depends on how reactive it is.

Metals above carbon

Reactive metals cannot be extracted by displacement using carbon. Instead, **electrolysis** is used. This is when an electric current splits up a compound into its elements. This uses a lot of energy, which is one of the reasons why these metals are expensive.

Unreactive metals

The least reactive metals, including gold and platinum, are so unreactive that they do not form compounds easily. This means that they are found pure in the crust. They are expensive because they are rare.

Most iron that is extracted is converted to steel

Very unreactive metals like platinum are often used for jewellery

Progress Check

1. Rubidium is more reactive than potassium. Which method is used to extract it?
2. Tin is less reactive than iron but more reactive than lead. What method is used to extract it?

Worked questions

a) Rhys added small pieces of zinc, calcium and magnesium to different boiling tubes containing dilute nitric acid. Look at the results table below and write the metals in order of reactivity below.

Metal	Observations
Zinc	Sank to the bottom, tiny bubbles appeared, stayed cold
Calcium	Lots of bubbles, got very hot
Magnesium	Floated and sank, bubbles, got warm

(2 marks)

Most reactive:	Magnesium
	Calcium
Least reactive:	Zinc

a) A question like this will often only have two marks available, because if you get the top and bottom answers correct, the middle one has to be correct. Questions where you draw lines to connect three boxes on the left with three on the right are similar: only two marks might be available.

b) Describe how Rhys would have made this a valid (fair) test. *(2 marks)*

He should always use the same type of acid. He should also use the same sized piece of metal each time.

b) It is important to remember that **fair testing** involves changing only one variable at a time. Many students confuse this with accuracy and reliability. One mark for a variable (name of acid, volume of acid, starting temperature of acid, etc.); and another mark for saying that it should be kept constant.

c) Here is a part of the reactivity series.

Most reactive:	potassium
	sodium
Least reactive:	lithium

Use this information to predict the products of the following reactions. If no reaction will occur, write **no reaction**. *(6 marks)*

potassium	+	lithium bromide	→	potassium bromide + lithium
Sodium	+	lithium nitrate	+	sodium nitrate + lithium
Lithium	+	potassium sulfate	+	no reaction

c) The first equation has been correctly answered, because potassium is more reactive than lithium, so it displaces it.

In the second reaction, the sodium is more reactive than the lithium, so the lithium is displaced.

In the final reaction, the lithium cannot displace the potassium.

Practice questions

1. This question is about displacement reactions. In a displacement reaction, a more reactive metal will displace a less reactive metal from a compound.

    a) Three metals, labelled **A**, **B**, and **C** were placed into solutions of **A sulfate**, **B sulfate** and **C sulfate**. The observations were recorded in the table below.

	Solution of A sulfate	Solution of B sulfate	Solution of C sulfate
Metal A	X	No change	Changed colour
Metal B	Changed colour	X	Powder appeared
Metal C	No change	No change	X

    Place these metals in order of reactivity. Explain why you have placed them in this order. *(4 marks)*

    b) Finish this word equation for the reaction between metal **A** and the solution of **C sulfate**.

    A + C sulfate → *(2 marks)*

    c) Metal A reacts with dilute hydrochloric acid to give small bubbles and a slight temperature rise. Suggest what will happen when metal B is added to dilute hydrochloric acid. *(2 marks)*

2. This question is about extracting metals. Tin ore was mined in the southwest of England for hundreds of years until the 20th century. Tin is a relatively unreactive metal, being placed below iron in the reactivity series, but above copper. Aluminium ore is mined in many countries, including Australia and Brazil. Aluminium is used in construction, cars, aeroplanes and cooking utensils.

    a) Tin can be extracted from tin oxide by heating with carbon. Suggest a word equation for this displacement reaction. *(2 marks)*

    b) Explain why carbon can be used to extract tin from its ore and why electrolysis is not needed. *(3 marks)*

    c) State the method used to extract aluminium. *(1 mark)*

    d) Explain why aluminium is more expensive than tin. *(2 marks)*

Learning Summary

After completing this chapter you should be able to:
- describe the structure of the Earth
- list some useful substances that can be extracted from the crust
- describe the rock cycle and how three types of rock are formed
- state the composition of the atmosphere and describe how human activity is affecting it.

42

14.1 Structure of the Earth

The structure of the Earth

Composition of the Earth

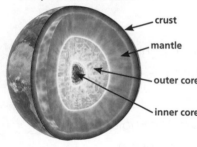

- crust
- mantle
- outer core
- inner core

The crust

The **crust** is the solid outer part of the Earth. Most of the crust is covered by the oceans. The crust provides us with many useful resources. The crust is made from huge sections called **tectonic plates**. These plates float on the mantle, which is moving very slowly, dragging them around over thousands and millions of years. The movements of the crust cause earthquakes and volcanoes.

The mantle

The **mantle** is hotter than the crust and is made from semi-solid rock, which can move very slowly.

The core

The **outer core** is made from very hot liquid iron and nickel, which is constantly moving. This movement causes the Earth's magnetic field. The inner core has the same composition but it is under greater pressure, so it is a solid. It is the hottest part of the Earth and is around the same temperature as the surface of the Sun.

Scientists have built up an understanding of what the inside of the Earth is made from, even though we have never drilled (or been!) below the crust. It would be impossible to journey into the mantle or deeper layers because of the heat and pressure. Scientists worked out the structure of the Earth by studying how different types of waves from earthquakes pass through the Earth.

Progress Check

1. Which is the coolest part of the Earth?
2. Which is the thinnest part of the Earth?
3. Which is the only truly liquid part of the Earth?

14.2 The Earth's crust

Composition of the crust

Element	Percentage (%) of the crust
Oxygen	46
Silicon	28
Aluminium	8
Iron	5
Calcium	4
Sodium	3
Potassium	3
Magnesium	2
Other elements	1

The table shows the percentage of some elements in the Earth's crust. The oxygen is present as part of compounds, mainly bonded to silicon, aluminium and iron.

Social and economic implications: useful resources from the crust

The crust contains a limited supply of fossil fuels (coal, crude oil and natural gas), which are extracted and processed to provide fuels for transport, power stations and domestic heating. Crude oil is also the starting point for making most plastics and many medicines and other useful chemicals.

Oil pumps

Metals are also extracted from the crust. Some are in very short supply, such as platinum, which is a valuable **catalyst** in reducing pollution from car exhausts.

Where resources are limited, it is sensible to recycle items to prevent waste, conserve **energy** and save money. **Recycling** plastics can help to reduce the amount of oil that must be extracted. Recycling used catalytic converters from old cars helps to provide platinum for new catalysts. Silicon is an abundant element in the crust (sand is mainly silicon oxide) so our demand for silicon for electronic devices is unlikely to mean that we will run out of it in the crust.

Make a three-dimensional model of the structure of the Earth, including labels to revise the key points.

Polymers

Polymers are long-chain molecules that are made from lots of smaller molecules added together, a bit like a chain of paperclips. They are made inside living cells (e.g. starch, cellulose and proteins) but can also be man-made (e.g. all plastics).

Plastics have a range of useful properties, so they can be used for a variety of different purposes. Here are some examples…

Bulletproof vest

Polymer	Used for...	Relevant properties
PET	Water bottles	Colourless, so you can see how clear the water is.
Poly(ethene)	Plastic bags and buckets	Strong, easy to mould.
PTFE	Non-stick coating for pans	Very low friction and very high melting point.
PVC	Gutters and window frames	Stiff, easy to mould.
Kevlar	Bulletproof vests	Incredibly strong.

Ceramics

Ceramics are made from chemicals found in the crust. They include porcelain and glass and are made by heating and cooling chemicals. As well as using ceramics for kitchenware, they can be used in high-performance cars (ceramic brake discs), in medicine (in bone repair surgery) and in the aerospace industry (nose cones for missiles and spacecraft).

Composites

A **composite** material is one that is made from two or more other substances in such a way that the new material has useful properties. For example, carbon fibre is a flexible woven fabric that is very strong. Epoxy resin is a brittle polymer. When combined, they form *carbon-fibre reinforced polymer*, which is a very stiff, strong and lightweight composite. This is used in racing cars, aeroplanes and sports equipment.

Write the names of some polymers on cards, make others and write properties on them and others that have uses of these polymers. Mix them up and see if you can then match the polymer with its properties and what it is used for.

Engine start button surrounded by carbon fibre reinforced plastic

Progress Check

1. Which is the most abundant metal in the Earth's crust?
2. Which rare metal is used in catalytic converters?
3. Give one use for carbon-fibre reinforced polymer.

14.3 The rock cycle

 44

Types of rock

There are three types of rock: igneous, sedimentary and metamorphic.

Igneous rock is made from molten rock, as a result of volcanic activity. It is very hard, which means that it is very difficult to scratch but it is good at scratching other substances. Igneous rock is made from crystals that are formed as the rock cools down. The slower the rock cools, the larger the crystals. Granite is formed deep underground (an

Basalt has very small crystals

Granite has very large crystals

intrusive rock), so it cools slowly and has large crystals. Basalt is formed on the surface (an **extrusive** rock), so it cools more quickly and has smaller crystals.

Sedimentary rock is made from tiny pieces of other rocks that settle out of slow moving water at the bottom of oceans and lakes. This insoluble material is called sediment. Over millions of years, the sediment is compressed and cemented (glued) together. Sedimentary rock is often in layers and may contain fossils. Limestone and sandstone are sedimentary rocks.

Sedimentary rock is often made from layers

A fossil in sedimentary rock

Metamorphic rock is made from either sedimentary or igneous rock when it is heated and compressed over millions of years. It may have layers of visible, interlocking crystals. For example, limestone can be converted to the metamorphic rock marble.

The rock cycle

The three types of rock are linked by the rock cycle, which is represented in the diagram below.

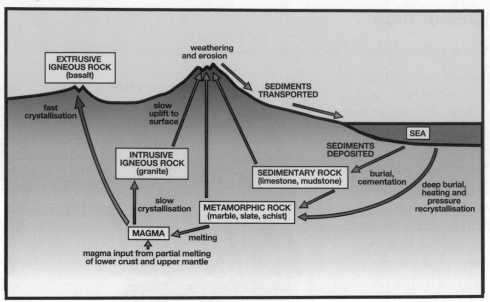

Igneous rocks are formed from volcanoes.

When the igneous rocks are on the surface, they can be worn away by the action of the weather.

The tiny pieces of rock are transported by rivers to the sea.

The sediment settles out of the slow moving water.

Sediments cement together as water is squeezed out.

Sedimentary rock is formed.

Movements in the tectonic plates cause this rock to be heated and compressed. This may form metamorphic rock.

Progress Check

1. Rhyolite is an igneous rock with very small crystals. Suggest how or where it was formed.
2. What type of rock is formed from tiny particles of other rocks at the bottom of the ocean?
3. Why could igneous rock never contain fossils?
4. What two conditions are required for the formation of metamorphic rock?

14.4 Composition of the atmosphere

45

The table and pie chart show the composition of dry air in the atmosphere today.

Gas	Percentage in dry air
Nitrogen	78%
Oxygen	21%
Argon	Nearly 1%
Carbon dioxide	0.04%

You can see from the table that there is a very small amount of carbon dioxide in the atmosphere. It is too small to be seen on the pie chart.

- Nitrogen
- Oxygen
- Argon
- Carbon dioxide

Social and economic implications: the carbon cycle

The amount of carbon dioxide in the atmosphere is very important because of the role it plays in the global climate. Several processes (natural and artificial) affect the movement of carbon atoms through different compounds and, together, these make the carbon cycle. Some of the processes have been colour coded and labelled. Understanding how humans are affecting these processes is important to help us minimise the negative impact that we have on the atmosphere.

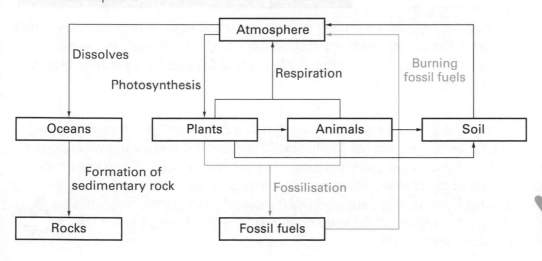

Use pictures, labels and pieces of string to make a bright display of the carbon cycle.

How are humans affecting the atmosphere?

One of the most important ways in which humans are affecting the atmosphere is by upsetting some of the processes in the carbon cycle. For example:

- Burning fossil fuels (coal, oil and gas) releases carbon dioxide that was trapped millions of years ago by photosynthesis.

- Cutting down trees (deforestation) prevents them from absorbing carbon dioxide by photosynthesis. This increases the amount of carbon dioxide in the atmosphere.

- Intensive farming of cattle causes the release of large amounts of methane from their waste gases.

Methane and carbon dioxide are both greenhouse gases. This means that they contribute to the **greenhouse effect**. The greenhouse effect describes how certain gases in the air absorb infra-red radiation given off by the warm Earth. This process helps to keep the Earth at a warm and stable temperature, which is important for all life on Earth.

Over the past few hundred years, human activity has increased the amount of greenhouse gases in the atmosphere so much that the average temperature of the Earth has increased. This is called global warming. Scientists predict that global warming will

speed up the melting of the polar ice caps, causing sea levels to continue to rise and disrupting weather patterns around the world. This is likely to cause flooding in coastal cities and crops are likely to fail in many countries around the world, leading to famine.

A few decades ago, scientists debated whether or not global warming was caused by human activity. Many scientists believed that there was enough evidence to support this theory but not everyone agreed. Since then, more data have become available and now almost all scientists agree that burning fossil fuels is causing global warming. Most of the scientists who did not agree before have changed their minds because of the new evidence.

Progress Check

1. Name two greenhouse gases.
2. Which gas makes up most of the atmosphere?
3. Which gas is absorbed from the atmosphere by photosynthesis?
4. Name three fossil fuels.

Worked questions

a) Match up the parts of the Earth with the correct statement. *(3 marks)*

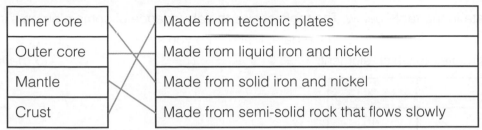

Inner core	Made from tectonic plates
Outer core	Made from liquid iron and nickel
Mantle	Made from solid iron and nickel
Crust	Made from semi-solid rock that flows slowly

b) Which is the most dense part of the Earth? Choose from the list above. *(1 mark)*

The inner core is the most dense part of the Earth.

c) Crude oil is a natural resource that is extracted from the Earth's crust. It is the starting point for a variety of chemicals, including petrol and diesel for cars, medicines and plastics. Explain why it is important to use crude oil reserves carefully and slowly and list some ways in which this can be done. *(3 marks)*

We should use it slowly because there is a limited amount of it in the crust and one day it will run out. This can be done by recycling plastics and using biofuels that are made from plants.

d) The properties of some materials are listed below. *(2 marks)*

Material	Properties
Glass	Brittle, transparent, stiff
Acrylic	Transparent, flexible
PTFE	White, flexible, slippery
Kevlar	Very strong yellow thread

Which of these materials would be best to use for making safety goggles for a science lab? Explain your answer.

Acrylic, because it is flexible as well as transparent, so you can see through it but it won't shatter if dropped.

a) There are usually three marks for this type of question because if you have three correct lines, the fourth line is obvious.

b) Note that, for a one-mark question, you don't usually need to bother explaining your answer unless the question specifically asks you to do this.

c) One mark is awarded for the idea that crude oil reserves are limited.

The second and third marks are awarded for describing ways to conserve crude oil. Mentioning ways to reduce energy consumption (such as turning off lights) would also be fine.

d) For this kind of question, one mark is awarded for a sensible answer. The second two marks are for justifying your choice using ideas about properties.

Practice questions

1. This question is about the Earth's crust.

   a) Look at the data in the table below. The table shows the abundance of some elements in the Earth's crust. Plot a pie chart of these data. Make sure you show which segment of your pie chart relates to which element. *(3 marks)*

Element	%
Oxygen	46
Silicon	28
Aluminium	8
Iron	5
Other elements	13

   b) The crust contains a number of fossil fuels. State two fossil fuels. *(2 marks)*

   c) Pure silicon is used for microchips in millions of electronic products that are sold every year. Pure silicon is very expensive. Silicon is found in the Earth's crust bonded to other elements. For example, sand is mainly silicon oxide. Suggest why pure silicon is so expensive. *(1 mark)*

   d) The crust is made from tectonic plates, which float on the mantle. When these plates move, they can have devastating effects on human populations. Describe some of these effects. *(2 marks)*

2. This question is about different types of rock.

   a) Peter is comparing some rock samples in a museum. He notices that one of them has very small crystals and one of them has very large crystals. He knows that both are igneous rocks. Suggest how each rock was formed. *(4 marks)*

   b) Peter finds a rock that has a fossil in it. What type of rock is he looking at? *(1 mark)*

   c) Limestone can be heated and compressed over millions of years and it becomes marble. What type of rock is marble? *(1 mark)*

3. This question is about how humans are affecting the atmosphere. Scientists and politicians around the world now accept that human activities are causing global warming due to the release of carbon dioxide and other greenhouse gases.

a) State how much carbon dioxide there is in the atmosphere. *(1 mark)*

b) Name two natural processes that remove carbon dioxide from the atmosphere. *(2 marks)*

c) Name one natural process that releases carbon dioxide into the atmosphere. *(1 mark)*

d) Explain how the following human activities increase the amount of carbon dioxide and methane in the atmosphere. *(3 marks)*

(i) Increasing population.

(ii) Deforestation.

(iii) Intensive farming.

e) Your carbon footprint is a measurement of the amount of carbon dioxide that is added to the atmosphere as a result of your lifestyle. Describe two changes that individual people in the United Kingdom can make to their lifestyle to reduce their carbon footprint. *(2 marks)*

f) A few decades ago, some scientists were unsure about whether global warming was part of a natural cycle or whether humans were causing it. Now, almost all scientists agree that human activity is causing global warming. Explain why most scientists now agree. *(1 mark)*

g) In recent years, there seems to have been more stories in the news about freak weather around the world. Examples include hurricanes and flooding. Some people blame this unusual weather on global warming. Do you think that this is justified? Explain your answer. *(2 marks)*

Learning Summary

After completing this chapter you should be able to:
- draw Sankey diagrams to describe how energy is transferred from one store to another and explain the conservation of energy
- evaluate the use of different energy resources
- calculate the energy used by an electrical appliance and the cost of this energy
- explain three ways in which thermal energy can be transferred and describe how to reduce heat loss.

15.1 Energy stores and transfers

It is hard to define what energy actually is, but it is much easier to describe some of the things that energy can do. Energy is useful when it moves from one energy store to another. We often call the movement of energy an energy transfer.

There are several ways that energy can be stored. These are:

The **kinetic store** (e.g. in a moving object) 	The **elastic store** (e.g. in a stretched rubber band) 
The **gravitational store** (e.g. in an object that has been lifted up) 	The **magnetic store** (e.g. in two separated magnets that are attracting each other) 
The **chemical store** (e.g. in a mixture of a fuel and oxygen) 	The **electrostatic store** (e.g. in two separated charged objects that are attracting each other) 
The **thermal store** (e.g. in a hot object) 	The **nuclear store** (e.g. in a radioactive atom) 

Scientists have argued for many years about the best way to explain what energy is. Even today, there is a lot of discussion between physics teachers about this!

Transferring energy

Energy is useful when it moves from one store to another. A mixture of wood and air on bonfire night is not useful in itself but by burning the wood we can move energy from the **chemical store** of the fuel and oxygen mixture into the **thermal store** of our bodies and warm ourselves up. We can represent this energy transfer with a diagram.

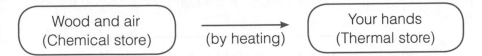

Wood and air (Chemical store) → (by heating) → Your hands (Thermal store)

In this example, energy moves by heating. There are other ways that energy can move from one store to another…

- Electrically (electrical current carries energy from one place to another).
- By radiation (light, sound, radio waves, etc. all carry energy from one place to another).
- Mechanically (when a force causes something to move).

These ways of moving energy from one store to another can often be used to label the arrows in energy diagrams. Here are some more examples of energy transfers.

A battery which is making an electric car move:

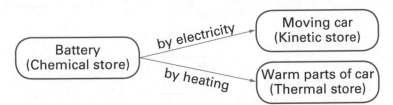

Battery (Chemical store) — by electricity → Moving car (Kinetic store); by heating → Warm parts of car (Thermal store)

We can see from this diagram that energy can move into two stores at the same time. Energy has a tendency to dissipate (spread out) and become less useful. Energy also tends to end up warming up the environment. Scientists and engineers spend a lot of time (and 'energy'!) trying to stop these two things from happening, which makes processes and **machines** more efficient.

Conservation of energy

The amount of energy that is transferred from one store to another can be measured. The units of energy are always joules (J). The way we calculate this depends on the transfer. Here are some examples.

When a model car travels up to the top of a hill and stops there, energy is transferred from its kinetic store to its gravitational store. The amount of energy in its kinetic store can be calculated using this equation…

Kinetic energy = ½ × **mass** × **velocity²**

(joules)   (kilograms) (metres per second)

Helpful tip: don't forget to square the velocity!

And the amount of energy in its gravitational store can be calculated using this equation…

Gravitational energy = **weight** × **height gained**

(joules)   (newtons) (metres)

When we do the calculations, we find that not all of the energy lost from the kinetic store goes into the gravitational store. Where has the 'missing' energy gone? The law of conservation of energy tells us that energy cannot be created or destroyed – it is simply transferred from one store to another. A **Sankey diagram** helps us to see where this energy has gone.

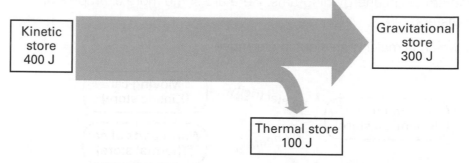

We can see that all of the energy transferred from the kinetic store of the car (400 joules) has gone somewhere and none has been lost. Most of it (300 joules) has gone into the gravitational store and the remaining 100 joules has been transferred by heating into the thermal store of the car and the surroundings.

If the car rolls back down the hill, the 300 joules that were in the gravitational store can now be moved into the car's kinetic store again. But remember that some of this energy will again be transferred to the thermal store of the surroundings, so there will be less energy available for the kinetic store. This means that the car will be slower at the bottom of the hill than it was before it started to climb it the first time.

Progress Check

1. Describe the energy transfer when you eat food that allows you to run a race.
2. Describe the energy transfer when you jump from a diving board.
3. Describe the energy transfer when you let go of a stretched elastic band.
4. Explain why a pendulum eventually stops swinging, using ideas about energy transfers.

15.2 Energy resources

There are many different types of **energy** resources. These are **not** the same as the stores of energy we considered in Topic 15.1! We can move energy from these resources in a way that is useful for us; for example, to make our televisions work and to keep our home warm in the winter:

- **fossil fuels**, like coal, oil and gas
- **biomass**, which means fuels that are produced from crops, like wood or biodiesel
- nuclear fuels, which can be used in special power stations
- water in reservoirs in the mountains, which can be used to run hydroelectric power stations
- wind, which can be used to drive wind turbines
- sunlight
- geothermal, which means hot rocks underground.

Energy from fuels

Fuels are substances that can be burned to move energy from their chemical store to a thermal store. This energy can often be used to generate electricity in a power station, because the thermal energy causes water to boil into steam and the steam drives a turbine which is connected to an electrical generator.

Fossil fuels are non-renewable, which means that they are being used faster than they can be produced and so one day they will run out. Fuels that are generated from growing crops are more likely to be **carbon neutral**, which means that they absorb as much carbon dioxide from the atmosphere when they are grown as they release when they are burned.

Energy from the Sun

Sunlight can be used to generate electricity, using photovoltaic (PV) cells. It can also be used to heat water for use in the home. Energy from the Sun causes weather, so wind and hydroelectric power are really ways of harnessing the energy from sunlight. So is biomass, because photosynthesis traps energy from the Sun.

Nuclear fuels

Nuclear fuels are obtained from the crust by mining and then purifying them. They are non-renewable, so one day we may run out. Generating electricity using nuclear power stations does not release any carbon dioxide, so it will not contribute to global warming. However, the radioactive waste produced must be stored for hundreds or thousands of years.

Cut out images of different energy resources and sort them into renewable and non-renewable. Then put them in order, in your opinion, of the most environmentally friendly through to the least environmentally friendly.

1. Name three renewable energy resources.
2. Name three non-renewable energy resources.

Progress Check

15.3 Energy from food

All animals obtain energy for their life processes from food. As always, energy is measured in joules (J). Because the joule is a small quantity of energy, sometimes larger units called kilojoules (kJ) are used. There are 1000 joules in one kilojoule. An old-fashioned unit of energy called the calorie (and kilocalorie, kcal) is also used on food labels.

When an animal (like you!) eats food and respires, energy is transferred from the chemical store of the food to the thermal and kinetic stores of the animal.

Chemical stores of energy in food

There are two main types of compound in foods that act as energy stores: fats and carbohydrates. Foods high in fats include…

- butter and margarine
- oils and fried food.

Carbohydrates can either be complex (starch) or simple (sugars). For example:

High in complex carbohydrates	High in simple carbohydrates
• pasta • bread • rice • potatoes	• fruit • chocolate • sweets • sugary drinks

Measuring the amount of energy stored in a food

Many foods can be burned to release energy from their chemical store. This energy can be transferred to the thermal store of some water in a boiling tube. A food that has more energy will cause a bigger temperature rise in the water.

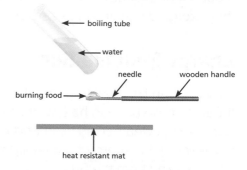

boiling tube

water

needle   wooden handle

burning food

heat resistant mat

To make this a valid (fair) test, you need to keep lots of things the same in this investigation:
- the volume of the water
- the mass of the food burned
- the distance between the food and the boiling tube
- the starting temperature of the water.

When evaluating this experiment, remember that lots of the energy released from the food is absorbed by the surroundings, not the water!

Remember that you learned more about the energy stored in foods in Topic 3.1.

Remind yourself about respiration by looking back to Topic 5.2.

Mime the stages of the investigation to determine which type of food contains the most energy.

Progress Check

1. Which food contains the greatest store of energy: celery or chocolate?
2. Is cheese high in fat, starch or sugar?

15.4 Power and appliances

A more powerful kettle will heat water for a cup of tea faster than a less powerful kettle. The power rating of a kettle can be found on a label on the bottom. **Power** is measured in watts (W) or sometimes in kilowatts (kW). One kilowatt is 1000 watts.

Power is a measurement of the rate at which energy is transferred. One watt means that one joule of energy is transferred every second. A high power rating means that an appliance transfers more energy each second than a less powerful appliance. You can calculate the energy transferred using the following equation…

A powerful iron (left); a less powerful iron (right) will take longer to heat up

> **Energy transferred** = **power** × **time**
> (joules)   (watts)   (seconds)

So a 2 kW kettle that is switched on for 30 seconds transfers this much energy…

Energy transferred = **2000 × 30**   (remember that 2 kW is 2000 watts)
(joules)   = **60 000 joules** (or 60 kilojoules)

Paying for domestic electricity

We use a lot of electricity in our homes, so the units we use to calculate it are a bit different, even if the equation is actually the same…

> **Energy transferred** = **power** × **time**
> (kilowatt hours)   (watts)   (seconds)

So if you leave an electric heater on for 3 hours and it has a power rating of 1.5 kW, the amount of energy transferred is…

Energy transferred   =   1.5 × 3
(kilowatt hours)   =   4.5 kilowatt hours (kWh)

If we pay for our electricity at around 15 pence for every kilowatt hour, running this electric heater for 3 hours will cost this much…

Cost of electricity =   4.5 × 15
(pence)   =   67.5 pence

---

1. If both were the same design, which bulb would be brightest: a 40 W bulb or a 100 W bulb?
2. How much would it cost to run a 100 W light for 12 hours if the electricity costs 15 pence per kWh?

Progress Check

50 15.5 Machines and work

Energy and work

Energy is often transferred from one store to another when a force is applied to an object and that object moves. This process is sometimes called 'doing work' on an object. Here are some examples…

- A car engine provides the force that allows the car to move at a constant speed (the force acts against **friction**).
- An object is lifted up, which involves pushing or pulling against gravity.
- Pushing an object along a surface, where the force overcomes friction.

The amount of energy transferred can be calculated by the following equation…

| **Work done (or energy transferred)** (joules) | = | **force** (newtons) | × | **distance moved in the direction of the force** (metres) |

For example, if a steady force of 20 newtons is used to push an object up a slope and that object moves a distance of 2.5 metres, the work done on the object is…

Work done $= 20 \times 2.5$
$= 50$ joules

It is worth pointing out that the energy in this process has probably come from a chemical store in the food eaten by the person doing the pushing. Some of the energy has been moved into the gravitational store of the object (it is now higher up the slope than it was) but most of the energy has been transferred into the thermal store of the object and the slope because of the friction between them.

Machines

A **machine** is something which makes use of the fact that, for a given amount of energy transferred, the smaller the distance moved, the bigger the force that is applied. The simplest type of machine is a lever, like using a long screwdriver to open a can of paint.

A small force is applied to the handle of the screwdriver over quite a large distance (a few centimetres in this case). The tip of the screwdriver moves only a very small distance but exerts a large force on the lid of the paint can, so it pushes it off easily. The work done by the person's hand is the same amount of energy as the work done on the lid of the can; as always, energy is conserved.

Use a force meter (newtonmeter) to pull some objects along surfaces or up slopes, keeping the force applied constant. Measure the distance you have pulled the objects and then calculate the work you have done on them.

Progress Check

1. The handle of a screwdriver is pressed down with a force of 20 N and moves a distance of 0.1 m. How much work was done on the screwdriver handle?
2. If the tip of the screwdriver moves by 0.01 m, what force does it exert on the lid of the can?

15.6 Temperature and heat energy

Heating and cooling

Energy moves from the thermal store of a hotter object (like this hot mug of tea) to the thermal store of a cooler object, or to the thermal store of the surroundings if they are cooler than the object. This is called cooling. The overall effect of this transfer of energy is to reduce the temperature difference between the objects.

When a cold ice cube is placed in warm surroundings, energy moves from the thermal store of the surroundings to the thermal store of the ice cube. The ice cube warms up and eventually melts.

You can also transfer energy to the thermal store of an object by rubbing it. Friction is a force that opposes (works against) movement. When friction acts between two moving objects, energy is transferred from the kinetic store of the moving objects to the thermal store of the objects.

Conduction

Conduction is a type of heat transfer when the energy is passed between particles without them moving from one place to another. Vibrating particles bump into their neighbours and cause them to vibrate, transferring the energy through a substance.

Conduction occurs best in solids because the particles are touching their neighbours. Metals are very good **conductors** of heat. Most non-metals are poor conductors of heat. Conduction can also occur in liquids, although convection is more important in liquids. Gases are very poor conductors of heat.

Convection

Convection occurs when hot particles move to a different place and take energy with them. Convection only occurs in liquids and gases because the particles of solids cannot flow. Convection currents are set up because hot liquids and gases expand and become less dense. This makes them rise and causes cooler parts of the liquid to sink. These current are most noticeable when the liquid or gas is heated from below, as in this pan on a stove.

Radiation

Radiation is the third method of heat transfer. Hot objects give off **infra-red** radiation, which is like light that we cannot see. Infra-red radiation is a type of electromagnetic radiation which has a wavelength that is longer than visible light. Infra-red radiation travels at the speed of light and it can travel through a **vacuum** and through most gases. Infra-red can be reflected by silver and white surfaces and it is absorbed by black surfaces.

Preventing heat transfer

Houses are insulated in many ways. Building the walls in two layers of brick with a vertical cavity (gap) between them reduces conduction through the walls, because air is poor conductor. Putting foam insulation into the cavity and into the loft space reduces heat loss by convection because it traps air pockets and prevents convection currents. If one side of the foam insulation is a shiny reflective foil, this reduces heat loss by radiation.

This thermal image shows where most of the heat is lost from this house. The hottest areas (white) are the windows, so double glazing would be a good idea. Blue is the coolest area, so the loft and roof are quite well insulated.

Work with some friends to devise a role play to demonstrate the three methods of heat transfer.

Progress Check

1. When you stir a hot drink with a metal spoon, the handle eventually heats up. Why?
2. When you hold your hand above a burning candle, it feels warm. Why?
3. What colour T-shirt should you wear to keep cool on a hot sunny day?

Worked questions

a) Hot water is poured from a kettle into a mug and left to stand for 20 minutes. At the same time, a cold drink is taken out of the fridge and it is also left to stand for 20 minutes. Describe the energy changes taking place during this time. *(2 marks)*

Energy moves from the thermal store of the hot drink and into the thermal store of the room. Energy moves from the thermal store of the room and into the thermal store of the cold drink.

b) A car transfers 1000 kJ every second when it accelerates. 100 kJ is transferred from the chemical store in the fuel to the kinetic store of the car and the rest is transferred to the thermal store of the surroundings. Draw a labelled Sankey diagram to represent the energy changes when the car is accelerating. *(3 marks)*

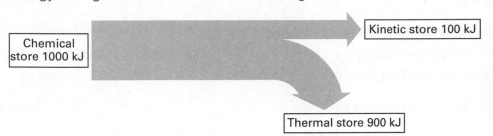

Kinetic store 100 kJ

Chemical store 1000 kJ

Thermal store 900 kJ

c) The driver brakes and the car slows down and stops. Describe the energy changes that take place during this time. *(2 marks)*

The energy moves from the kinetic store of the car to the thermal store of the surroundings.

d) Calculate the amount of energy transferred by a 1.5 kW electric heater that has been left on for 10 seconds. Include the units in your answer. *(2 marks)*

Energy transferred (J) = power (W) × time (s)

$Energy = 1500 \times 10$

$= 15\,000\,J$

a) Explaining that energy moves from a hot substance to a colder substance (like the surroundings) is the key idea here. Coldness doesn't flow anywhere.

b) For a Sankey diagram like this, one mark will be for the correctly sized arrows. Another mark will be for the correct labels. The third mark will be for the numbers and units.

c) You need to correctly identify the energy stores in a question like this.

d) The equation needs you to convert the power from kilowatts into watts before you can do the calculation. 1.5 kW is 1500 watts.

Practice questions

1. This question is about energy changes in the kitchen.

    a) Rahima is making a sauce. She uses the gas hob to melt the butter. Describe the
       energy changes that are taking place. *(2 marks)*

    b) Rahima uses the oven to roast a chicken for 1.5 hours. The oven has a power rating
       of 2 kilowatts. Calculate how much it costs Rahima to pay for the energy needed to
       roast her chicken. Electricity costs 15 pence per kWh.

       Energy transferred (kWh) = power (kW) × time (hours) *(2 marks)*

    c) Whilst cooking the chicken, Rahima wraps some parts of the chicken in shiny
       aluminium foil to prevent them from burning. Use ideas about heat transfers to explain
       why she does this. *(2 marks)*

    d) To cook the vegetables, Rahima boils some water in a kettle and then pours it into
       the pan on the hob before adding the vegetables. The kettle transfers 45,000 joules
       in 30 seconds. Calculate the power rating of the kettle.

       Energy transferred (J) = power (W) × time (s) *(2 marks)*

    e) Rahima's brother, Iman, suggests to her that she should put the lid on the pan
       whilst cooking the vegetables 'to keep the heat in'. Explain how heat is lost from a
       pan without a lid and why putting a lid on it helps to prevent it cooling down. *(2 marks)*

2. This question is about keep warm during and after a triathlon. A triathlon is a race
    that involves swimming, cycling and running.

    a) During the swimming stage of the triathlon, the main type of heat loss is by conduction.
       Describe how the water transfers heat away from the swimmer's body by
       conduction. *(2 marks)*

    b) Explain why heat loss by conduction is less of a concern when cycling and
       running. *(1 mark)*

    c) During the cycling and running stages of the event, the competitors often wear
       clothing that traps a layer of air close to the skin. Identify the main type of heat transfer
       occurring during cycling and running and explain how their clothing keeps the
       competitors warm. *(2 marks)*

    d) At the end of the event, it is common for shiny foil blankets to be given to competitors
       so that they can wrap the blanket around their body. Identify the type of heat transfer
       that this attempts to reduce and explain why the blankets are more effective at doing
       this than a fleece jumper. *(2 marks)*

3. This question is about the future of the energy supply in the United Kingdom. For many years, the vast majority of the energy supplied to UK homes has come from non-renewable resources.

a) What is meant by the term **non-renewable** when applied to energy resources? *(1 mark)*

b) The table below shows some data about which energy resources are used to generate electricity for UK homes and businesses. Plot a suitable graph or chart to represent these data. *(4 marks)*

Energy resource	Quantity of electricity generated / TWh
Gas	160
Coal	130
Nuclear power	70
Wind	20
Hydroelectricity	10

c) In 2013, the UK government announced a very large project that would involve building new nuclear reactors at Hinckley Point in Somerset and at Sizewell in Sussex. Suggest two reasons why the government wants to reduce the amount of electricity that is generated by gas and coal. *(2 marks)*

d) Suggest two reasons why some people have opposed the development of new nuclear power stations. *(2 marks)*

e) The UK is the sixth largest producer of electricity from wind power in the world. Evaluate the advantages and disadvantage of wind power as a way of producing electricity for the UK today and in the future. *(4 marks)*

4. This question is about the energy stored in foods.

a) Name two types of chemical that act as significant stores of energy in the food we eat. *(2 marks)*

b) For each of these types of chemical, give an example of a food that is high in them. *(2 marks)*

c) Describe an experiment that will allow you to measure and compare the amount of energy stored in samples of different foods. Include the following in your answer.

- Apparatus and diagram
- What you will measure
- How you will process your results

- What you will change
- What you will keep the same
- How you will conclude which food contains the largest store of energy *(6 marks)*

Learning Summary

After completing this chapter you should be able to:
- draw distance-time graphs to represent a journey
- calculate the average speed of a moving object
- draw force diagrams to represent balanced and unbalanced forces.

52

16.1 Speed, distance and time

Speed is how fast something is going. It is a measurement of how far something travels in a specific time. Average speed is calculated using the following equation.

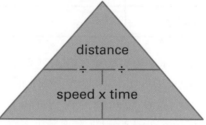

$$\text{Average speed} = \frac{\text{distance travelled (metres, m)}}{\text{time taken (seconds, s)}}$$
(metres per second, m/s)

You can find out how to use equation pyramids like this in **Maths skills for science** on page 19.2.

For example, if it takes a swimmer 40 seconds to swim 50 metres, her average speed is calculated like this...

Average speed $= \dfrac{50\text{ m}}{40\text{ s}}$ (Notice that, for calculations involving speed, it is much better to have the time in seconds than in minutes)

$= 1.25\text{ m/s}$

If the swimmer could keep this speed up for a 200 m race, how long would it take her to finish?

Time $= \dfrac{\text{distance}}{\text{speed}}$

Time $= \dfrac{200\text{ m}}{1.25\text{ m/s}}$

Time = 160 s. You can convert this to minutes by dividing by 60, to give 2 minutes and 40 seconds.

Distance–time graphs

A distance–time graph tells the story of a journey. For example, if you leave your house and walk to school, your journey might involve a quick stop at the shop to pick up something to eat. Then you realise you are late, so you run the rest of the way. The distance–time graph for this journey might look like this.

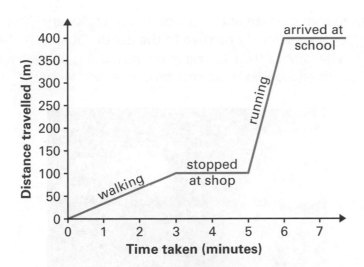

From this graph, we can see that it took you 3 minutes to get to the shop. Notice that the sloping line on this graph means that something is moving. Because it is a straight sloping line, it means that you were walking at a constant (steady) speed. You stopped at the shop for 2 minutes (between minute 3 and minute 5 of your journey). Then you ran to school, which took you 1 minute. Notice that the line is steeper when you move faster.

In total, you moved 400 metres in 6 minutes on this journey. So your average speed is calculated like this…

First, we need to convert 6 minutes into seconds, so multiply 6 by 60 to give 360 seconds.

$$\text{Average speed} = \frac{400\ m}{360\ s}$$

Average speed = 1.1 m/s

Use a measured distance like an athletics track and time yourself walking and then running this distance. Work out your speed. Remember to use metres and seconds.

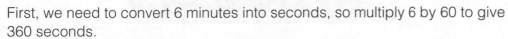

1. A car travelled 1000 m in 40 seconds. What speed was it travelling?
2. Looking at the graph of your journey to school, how far was it from the shop to school?
3. How long were you running for, between the shop and school, in seconds?
4. What was your speed whilst running?

Progress Check

53

16.2 Relative motion

We often describe something as stationary (standing still) or moving. But these terms are **relative**. This means that we need to know what we are comparing them to.

You may think that your house and your school are stationary. What we really mean is that they are stationary **relative to the Earth**. Of course, the Earth is moving all the time, spinning on its axis and also moving through space as it orbits the Sun. Even the Sun is moving through space as part of the Milky Way galaxy.

You will learn more about the movement of the Earth and the other planets in Topic 19.2.

If you sit on a train and look out of the window at the ground, houses and trees, you can see that you are moving very fast. But if your train overtakes a slower train travelling in the same direction and that is the only thing you can see out of the window, it looks as if your train is moving slowly. This is because your speed **relative to the slower train** is much lower.

If your fast train passes a train going in the other direction, your **relative speed** is very fast indeed. This explains why head-on collisions in cars are so dangerous. If two cars travelling at 30 m/s crash into each other head on, this would be equivalent to one car crashing into a solid brick wall at 60 m/s.

Progress Check

1. Calculate the relative speed of the impact when two cars crash head-on, if one is going at 20 m/s and one is going at 10 m/s.
2. If your car travels at 30 m/s and overtakes another car travelling in the same direction at 25 m/s, what is your speed relative to the slower car?

16.3 Forces

A force is a push or a pull. It acts between two objects. The size of a force is measured in newtons (N). The direction that a force acts in is also of great importance, so wo represent forces in diagrams with arrows to show this. The size of the arrow represents the size of the force.

It is important to consider all the forces acting on an object. For example, gravity acts on all objects, all of the time. Gravity even acts in space but it gets much weaker the further you go from a planet or star (like the Sun).

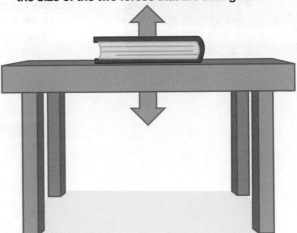

A book resting on a table; the arrows show the size of the two forces that are acting on it

Friction

Friction is a force that acts against a moving object or prevents a stationary object from starting to move if a force is applied to it. Sometimes friction is useful, such as in the brakes in a car, and between the tyres and the road. Sometimes friction is a nuisance, such as between the moving parts in a car engine, so we use lubricants (e.g. oil) to reduce friction.

What can forces do?

When a force acts on an object, it can do one or more of the following things…

* Change its shape; for example, when you squash a spring or stand on a football.
* Make it speed up or slow down; for example, when you throw or catch a ball.
* Make it change direction; for example, when a car turns a corner.
* Make it spin.

Balanced and unbalanced forces

Look around you and list all the examples of forces that you can see acting on different objects. Which are bigger and which are smaller?

When the forces acting on an object balance each other out, there is no **resultant** (overall) **force**. Here are two examples.

The two forces acting are balanced, so the resultant force is zero. The bowl stays stationary.

All the forces are balanced, so the resultant force is zero. The car carries on moving at a constant speed in a constant direction.

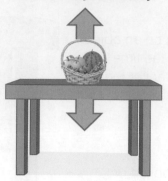

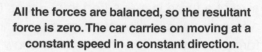

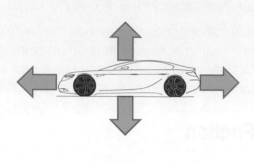

Remember that, if the resultant force is zero, one of two things will happen…
- A stationary object stays stationary.
- A moving object carries on moving, at a constant speed and in a constant direction.

If the forces are unbalanced then…
- A stationary object will change shape or start moving (or both).
- A moving object will change shape, change direction, speed up or slow down (or a combination of these).

Here are some examples…

The forces acting vertically are balanced but there is a resultant force of 800N acting on the car, making it accelerate (speed up).

The forces acting vertically are balanced but the engine is no longer providing a forward force and the driver has applied the brakes. The resultant force acts to slow the car down.

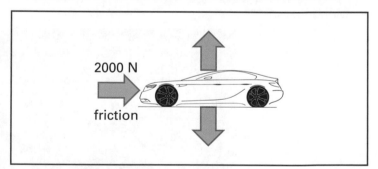

200 N
friction

1000 N
force from engine

2000 N
friction

It is important to remember that, when something is in orbit in space, around a star (like the Sun) or around a planet, there is a force acting on it. This force is gravity and it causes the object in orbit to constantly change direction, making it move in a circle. Without this force acting, the object would fly off into space in a straight line.

Force arrow diagrams

Sometimes, forces can be difficult to see but if we think about the changes that are taking place, this can help us to identify the forces that are acting.

Accelerating, so the drag must be less than the weight of the skydiver

drag ⬆ 50 N

weight ⬇ 700 N

Decelerating, so the drag from the parachute must be greater than her weight

drag ⬆ 1500 N

weight ⬇ 700 N

Constant speed, so the two forces must now be equal

drag ⬆ 700 N

weight ⬇ 700 N

Slowing down at the instant she touches the ground, so the reaction force from the ground must be more than her weight

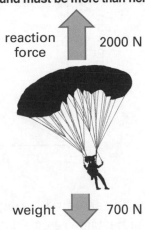

reaction force ⬆ 2000 N

weight ⬇ 700 N

Take a walk around and describe the forces that are acting on you at different times during your journey. Evaluate whether the forces are balanced or unbalanced at different times.

1. Calculate the size and direction of the resultant force in the first diagram of the skydiver when she is accelerating.
2. Calculate the size and direction of the resultant force when she has just opened her parachute.
3. Calculate the size of the resultant force when she is travelling at a constant speed.
4. Does she slow down faster when she opens her parachute or when she reaches the ground? Explain your answer.

Progress Check

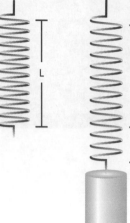

16.4 Hooke's law

It is easy to investigate the way that a force affects a spring. Here is the method…

1. Attach the top of a spring to a secure place so that it cannot come loose and so that, if the masses fall off, they will not cause injury or damage.
2. Measure the length of the spring, L.
3. Add a small mass to the bottom of the spring. Measure the length of the spring again and then subtract L from it, to find the **extension**, x.
4. Convert the mass to a force (weight) by multiplying the mass in kilograms by 10.
5. Repeat with other masses but be sure not to apply so much force to the spring that it deforms.
6. Plot a graph of force (x axis) against extension (y axis) and draw a line of best fit.

At each measurement, the forces on the spring are balanced. The weight of the mass on the spring is balanced by the tension in the spring. Here is a typical graph obtained from a Hooke's law experiment.

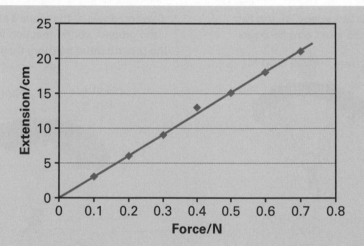

We can also see from the graph that there is one point that does not fit on the line of best fit. We could call this an **outlier** or an **anomalous result**. It would be a good idea to repeat this result.

Note that the independent variable (the one we changed) goes on the x-axis and the dependent variable (the outcome) goes on the y-axis.

The graph shows us that as the force increases, so does the extension. The line is straight and goes through the origin, so this is an example of a directly proportional relationship. This means that, as we double the force, the extension also doubles. For example, with a force of 0.3 N, the extension is 9 cm. When the force is doubled to 0.6 N, the extension is 18 cm.

When doing this experiment, if too large a force is used, the **elastic limit** of the spring is reached, and it will no longer return to its original length. At this point, the line on the graph starts to curve. The spring has now been deformed and the change in its shape is now **plastic** (permanent), rather than elastic.

Progress Check

1. When the force applied to the spring was 0.5 N, what was the extension?
2. Why must the graph go through the origin in this investigation?

16.5 Moments

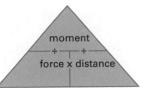

Sometimes, forces have a turning effect on an object. This effect is called a **moment**.

The size of a moment can be calculated using the following equation…

> **Moment = force × perpendicular distance from pivot**
> (newton metres, Nm)   (newtons, N)   (metres, m)

moment
÷
force x distance

If the moments on an object are unbalanced, the object will either…
- start to spin if it is stationary
- spin faster or spin slower if it was already spinning.

If the moments on an object are balanced, the object will either…
- stay stationary
- carry on spinning at the same speed in the same direction.

A familiar example of moments is a see-saw which is stationary and balanced.

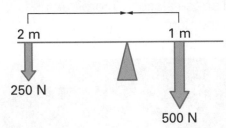

The anti-clockwise moment caused by the mass on the left of the see-saw is 2 × 250 = 500 Nm. The clockwise moment caused by the mass on the right of the pivot is 1 × 500 = 500 Nm. So the see-saw is balanced.

You can use the idea of balanced moments to do calculations to find a missing force or missing distance, because…

$$\text{Force}_1 \times \text{distance}_1 = \text{force}_2 \times \text{distance}_2$$

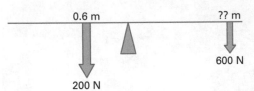

For example, calculate the distance from the pivot at which the 650 N force must be applied to balance the moment of the 200 N force.

$$200 \times 0.6 = 600 \times ??$$

$$?? = \frac{200 \times 0.6}{600}$$

$$= 0.2 \text{ m}$$

Make a see-saw and explore the principle of balanced moments by trying to balance a heavy weight near to the pivot with a lighter weight further from the pivot.

1. If you weigh less than your sister and you are both sitting on opposite ends of the see-saw, who must move and where, to balance the see-saw?
2. Calculate the moment of a 2000 N force that is applied 3 m from the pivot.

Progress Check

16.6 Non-contact forces

Many everyday forces that we can study are **contact forces**. This means that they act between objects that are touching.

Some forces act between objects that are not touching, so we call them **non-contact forces**.

Gravity

Gravity acts between any two objects. The size of the gravitational attraction between two objects depends on the mass of both objects and also the distance between them. You are attracted to this book but your mass and the mass of the book are so small that the attraction is tiny. The mass of the planet is very large, so the gravity between you and the planet is much larger and enough to keep your feet on the ground.

For an astronaut, however, the distance between them and the Earth is very large, so the gravitational pull on them is much smaller than it is on Earth. They appear to be weightless if this force is small enough. But gravity does act in space. It is what keeps satellites and space stations in orbit around Earth and it keeps the Earth in orbit around the Sun.

Magnetism

Two magnets attract if their opposite poles are near to each other. They repel if their North poles are facing each other (or their South poles).

Static electricity

When an electrical charge has built up on an object, it can exert a non-contact force on another object. If that object has the same charge, the two objects **repel** (push apart). If the other object has the opposite charge, the objects attract.

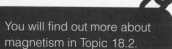
Space is explored in more detail in Topic 19.1.

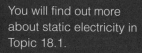
You will find out more about magnetism in Topic 18.2.

You will find out more about static electricity in Topic 18.1.

Take an inflated balloon and rub it on your clothing. See if you can experience the non-contact force as you hold the balloon near to your hair.

Progress Check

1. Give three examples of non-contact forces.
2. On which two factors does the size of a gravitational force depend?

16.7 Pressure in fluids

Fluids are substances that can flow, so this includes liquids and gases.

Pressure is calculated using this formula.

(newtons per metre squared, N/m²) Pressure = $\dfrac{\text{force}}{\text{area}}$ $\dfrac{\text{(newtons, N)}}{\text{(metres squared, m}^2\text{)}}$

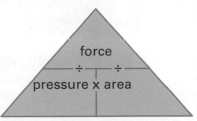

force ÷ ÷ pressure x area

Pressure can be increased by decreasing the area over which a force acts. Wearing studded football boots gives you more grip because the increased pressure pushes the studs into the soft ground.

Pressure can be decreased by increasing the area, which is how snow shoes stop you sinking into deep snow.

Air pressure

The air exerts a pressure on us which is surprisingly high. It is equivalent to a 1 kilogram mass resting on each square centimetre of our skin! We don't notice this because we have evolved to live in this atmospheric pressure and the pressure inside our bodies is the same. Air pressure comes from the weight of the air above us, right up to the edge of the atmosphere. This explains why the air pressure is less when you are higher up, on a mountain or in an aeroplane.

Water pressure, floating and sinking

Water pressure increases with depth, as the weight of the liquid above you increases.

When an object is placed into a liquid, it displaces some of the liquid. This can be seen if you fill a bowl right to the brim with water and then put something in it. Water overflows from the bowl.

Any object in a fluid experiences an upward force called **upthrust**. The size of this upthrust is equal to the weight of the liquid that the object displaces. If the upthrust is greater than the weight of the object, the object floats. If the weight of the object is greater than the upthrust, the object sinks.

Because the amount of water displaced depends on the volume of the object, and its weight depends on its mass, it is useful to consider the **density** of the object to help us to predict if something will float or sink. Density is given by this equation…

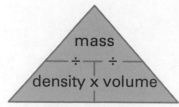

(kilograms per metre cubed)

$$\text{Density} = \frac{\text{mass (kilograms, kg)}}{\text{volume (metres cubed, m}^3)}$$

If something has a low mass for a given volume, it has a low density. This means it will displace a large volume of water and will have a large upthrust. It is likely to float.

Ships are designed with a shape that displaces a large volume of water for their weight, so that they can float even if they are made from steel, a very dense metal.

These same ideas explain floating in gases too. Hot-air balloons and helium balloons have a large volume and a low mass, so they displace a large volume of cold air and thus their upthrust is greater than their weight.

Use a large bowl of water and some household objects to investigate floating and sinking. Draw a force diagram for each object, showing clearly whether it floats or sinks, and why.

Progress Check

1. When you push on a drawing pin, the same force is exerted on your finger as on the board but it goes into the noticeboard, not into your finger. Explain why.
2. Explain why a helium balloon will expand when it rises up into the atmosphere.

Worked questions

a) Look at the distance time graph below for a pupil in a PE lesson. The time starts from the bell at the beginning of the lesson. *(1 mark)*

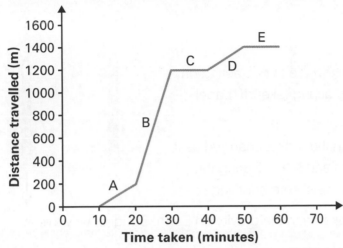

How long did the pupil spend getting changed at the start of the lesson?

Ten minutes

b) Section A on the graph is when the pupils were doing a warm up. Section B represents the race that the pupils ran. Calculate the speed of the pupil during this race. *(3 marks)*

Distance travelled = 1000 m.

Time taken = 10 minutes.

$$Speed = \frac{distance}{time}$$

$$= \frac{1000 \ m}{10 \ min}$$

$$= 100 \ m/min$$

c) Compare the graph at sections B and D. Suggest what the pupil was doing during section D. Explain your answer. *(2 marks)*

She was doing the cool down because the line at D is less steep than at B, so she was moving more slowly than during the race.

a) Ten minutes is the correct answer. This demonstrates that the student understands that for the first ten minutes the pupil did not move anywhere.

b) It is very common for students to make mistakes on this kind of question and calculate the average speed during the first 30 minutes of the lesson, not the speed of the pupil during the section marked B. During the section marked B, the pupil ran 1000 m (the difference between 200 m and 1200 m) during a period of 10 minutes (the difference between 20 and 30 minutes).

c) We can see that when the question asks you to **explain** something, you should generally use the word **because** in your answer.

Practice questions

1. This question is about the NASA Mars Rover *Curiosity*, which launched in 2011 and landed on Mars in 2012. It was taken to Mars by a rocket.

   a) Draw a force arrow to show the force acting on the rocket from the ground at the moment before the rocket takes off.

   (2 marks)

   b) Draw another diagram to show the forces acting on the rocket as it accelerates during take-off. Label the forces.

   (2 marks)

   c) Describe how the forces on the rocket changed as it travelled between the Earth and Mars. During this stage of the journey, the rockets were not firing.

   (2 marks)

   d) The gravity on Mars is weaker than the gravity on Earth. Suggest why this is.　*(1 mark)*

2. This question is about Formula 1 racing cars during a race.

   a) Look at the images below. Identify the pictures in which the forces acting on the car are balanced.

   (2 marks)

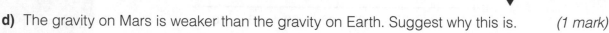
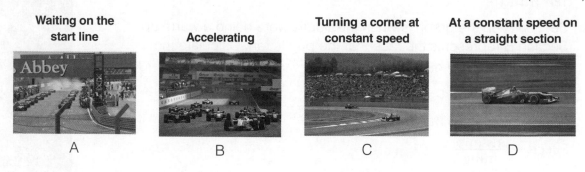

| Waiting on the start line | Accelerating | Turning a corner at constant speed | At a constant speed on a straight section |
| A | B | C | D |

   b) A Formula 1 race car will typically do a lap of a 6 km track in 1 minute and 30 seconds. Calculate the average speed of the car during this lap, in metres per second.　*(3 marks)*

   $$\text{Speed (m/s)} = \frac{\text{distance (m)}}{\text{time (s)}}$$

   c) The maximum speed of the car during a race may be as high as 100 m/s. When the car is travelling at this speed, calculate the distance that it would travel during 20 seconds.

   (2 marks)

   d) Each tyre of a Formula 1 car has an area of 0.1 m² that is in contact with the ground. There are four tyres on a car. The car weighs 6500 N. Calculate the pressure that the tyres exert on the ground.

   (3 marks)

   Pressure (N/m²) = force (N) / area (m²)

   e) The tyres are inflated to a pressure of 250,000 N/m². Assuming that the area of the inside of a Formula 1 tyre is 1.5 m², work out the total force that is acting on the inside of the tyre by the gas inside.

   (3 marks)

After completing this chapter you should be able to:
- describe the similarities and differences between waves in water, sound and light
- describe reflection, refraction and dispersion
- explain how we hear and see.

Learning Summary

17.1 An introduction to waves 🎧 59

Waves are regular disturbances that carry energy. There are waves that we can see, like waves in water, but also others that we can't see, like light and sound. Waves can be grouped into two types which can be shown using a slinky spring:

Type of wave	Image	Example
Transverse		Light and water
Longitudinal		Sound

Longitudinal waves are started by a movement in the direction of the wave (so left and right if the waves moves across the page to the right). Transverse waves are started by a movement at right angles to the movement (so up and down).

Some waves require a medium of particles (a solid, liquid or gas) to travel through. For example, sound waves cannot travel in the vacuum of space. Other waves don't require a medium. So light can travel in space.

Sound waves are longitudinal

Light waves are transverse

Use a slinky spring or length of string to make transverse or longitudinal waves. Can you increase the amplitude or frequency?

Wave diagrams

We show both transverse and longitudinal waves in the same way:

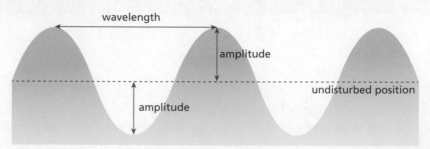

The **wavelength** is the horizontal distance between the same two points on any wave. This can be from the two crests of a wave (as shown in the diagram), the two troughs (bottom points) or any other two identical points. Wavelengths are distances and so measured in metres, although many are much smaller than this.

The **amplitude** is the vertical distance that the wave travels from its starting position; so from the middle to the crest or the middle to the trough. All waves transfer energy and the greater the amplitude the greater the energy it transfers.

The **frequency** of a wave is the number of complete waves that are produced or pass any point in a second. Frequencies are measured in hertz (Hz).

Adding and subtracting waves

When waves of the same frequency meet each other, they can either add together or cancel each other out. If the waves are in phase (their crests and troughs match up) then their amplitude will double. If the crest of one meets the trough of another, they will cancel out, as shown here.

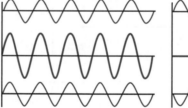

We hear the sea when we listen to a shell because tiny sound waves from the surrounding environment echo inside the shell and amplify the sound.

> Use a short length of string to model waves on a piece of paper. Can you double the frequency without changing the amplitude? Can you now do this the other way around?

Progress Check

1. State the two types of waves.
2. Describe the ways in which these waves are formed.
3. Explain why no one can hear you scream in space.

17.2 Sound waves

The vibrations of the diaphragm of this loudspeaker cause sound waves of vibrations through the air

Sounds are caused by vibrations and are actually waves of vibrating particles. Our voices are caused by air from our lungs which vibrates our vocal cords, sending waves of vibrations through the air. We hear them when they bump into and vibrate our ear drums.

The areas where the waves of particles are close together are called **compression** and those where there are few particles are called areas of **rarefaction**.

Sound waves are longitudinal waves. This means that they can only travel through solids, liquids or gases. The arrangement of particles in these media affects how quickly waves travel in them.

Sound waves spread out as waves of vibrating particles

Medium	Diagram	Speed	Reason
Solid		Fastest	The particles are closest together
Liquid		Medium	The particles are not as close as in solids
Gas		Slowest	The particles are furthest apart
Vacuum		Zero	There are no particles to vibrate

Put an ear close to the edge of a table. Stretch out your arm as far from your ear as possible and scratch the table so gently you can hardly hear it. Then put your ear on the table and repeat the same scratching. Why can you hear it much louder now?

Research the auditory ranges of different animals on the internet. Record the maximum frequency that a range of different animals can hear. Plot your data as a bar chat. How do the maximum frequencies you've found compare with that of a human?

Measuring sound waves

Some people use the non-scientific term loudness instead of amplitude of a sound. Loudness is often measured in decibels. Some people also use the non-scientific term pitch instead of frequency for sound waves. The frequency of all waves including sound waves is measured in hertz.

The range of human hearing is called our **auditory range**. This is between 20 and 20 000 hertz. The auditory range of other animals extends beyond ours. Cats can hear up to 79 000 hertz and bats even higher to 200 000 hertz.

As you get older it is harder to hear quiet sounds (those with low amplitudes). This will happen faster if you often listen to loud music.

Reflection and echoes

Sound waves can reflect from surfaces. Flat surfaces reflect sounds better than rough ones. If sounds reflect back to you a short period of time after you first heard them, you will hear them again although a little quieter. This is an **echo**.

Foam soundproofing absorbs sound waves and reduces reflective echoes

A duck's quack will echo despite many claims to the contrary

The ear

The ear is your organ of hearing. It detects sound, allowing you to hear.

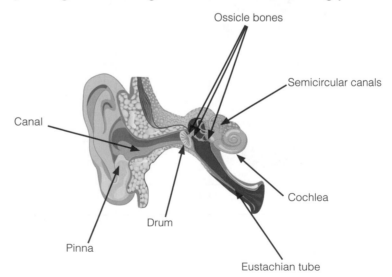

Ossicle bones

Semicircular canals

Canal

Cochlea

Drum

Pinna

Eustachian tube

You have two ears to help you to locate the direction of sounds. Cover both your eyes and one ear. Ask a friend to walk around you making sounds. Can you identify their direction as well with one ear as with two?

Pinna (ear flap)	Collects the sound vibrations from the surrounding air
Canal	Focuses the vibrations from your pinna on to your ear drum
Drum	Transfers the vibrations of the drum to the ossicle bones
Ossicle bones (hammer, anvil and stirrup)	Increases the pressure of the vibrations on the fluid in your cochlea
Cochlea	Contains tiny hairs in fluid which convert vibrations into electrical signals
Auditory nerve	Transfers the electrical signals to your brain
Semicircular canals	Filled with fluid to help you balance (does not help with hearing)
Eustachian tube	Connects the middle ear to the back of your nose to help equalise pressure in your ear (does not help with hearing)

Microphones detect sounds and convert them into electrical signals. They have a diaphragm inside them which detects the vibrations of sound waves just like the ear drum vibrates when we hear a sound.

Ultrasound

Ultrasound waves have frequencies above 20 000 hertz. This means they are beyond the range of human hearing. A pregnant woman always has an ultrasound scan to take images of her fetus to determine the size, position and health of the unborn baby. The waves are sent from a scanner into the mother's abdomen, where some reflect back (exactly like an echo) when they reach tissues or organs. This creates a digital image of the baby.

An ultrasound image of a baby at 4 months

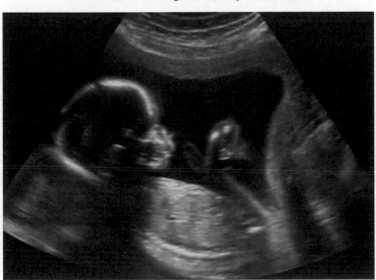

The fluid in your semi-circular canals acts like a spirit level and so helps you balance

Cochlea is the Latin word for snail because it looks a little like a snail shell

Changes in air pressure are common when flying or diving and are equalised by your Eustachian tube

Put your finger in the middle of the bottom part of your lobe. Move gently upwards following the channels in your pinna. Where does your finger finish and why is this important?

You will learn more about developing babies in Topic 4.3.

Ultrasound is also used to check internal organs of people who are not pregnant and can also be used to clean jewellery and watches.

Kidney stones are solid lumps that build up in kidneys causing great pain and can be identified by ultrasound

Scanning the thyroid of a man

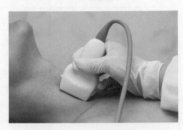

Ultrasound is also used in SONAR. SONAR stands for SOund Navigation And Ranging. Here boats or submarines send ultrasound waves out into the surrounding water. These reflect back from other objects, commonly the seabed or another boat or submarine. The time delay for the wave to return is used to calculate how far away the object is.

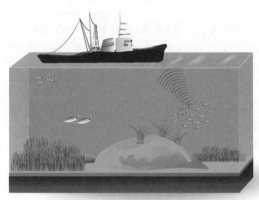

Use the corner of your exercise book to make a flicker book showing how SONAR works.

Infrasound

Infrasound waves have frequencies lower than 20 hertz. This means that they are below the range of human hearing. Many animals such as elephants and whales can communicate over huge distances using infrasound. Infrasound waves are also given off by volcanic eruptions.

Infrasound waves are given off by volcanic eruptions and are used by whales to communicate

Progress Check

1. State the scientific term for loudness.
2. Describe the function of the pinna.
3. Explain why sounds travel fastest in solids.

17.3 Light waves

Light waves are transverse and are the moving vibrations of energy and not particles like sound. So light waves can travel through a vacuum like space.

Light waves travel much faster than sound waves or those on water. This explains why we see a flash of lightning before we hear the sound made by thunder. In fact, light moves faster than anything else. The distances between objects in space, like those between us and our nearest stars, are so great that we measure them in light years. A light year is the distance light travels in one year and is approximately 10 000 000 000 000 000 metres. Waves of light take just over eight minutes to travel the 150 000 000 000 metres from the Sun to Earth.

You will learn more about space in Topic 19.1.

Stand still in the middle of a large open space like a park. Watch as a friend walks away from you banging two metal objects together, perhaps a spoon on a metal lid. How many paces do they have to take before you can see the movement of their hand before you can hear the sound? Why is this?

Drawing ray diagrams

Light always travels in straight lines. However, light rays move in transverse waves. We always draw them as a straight line that would pass through the middle of these waves. We draw one arrow on each wave to show the direction that the light is travelling. These drawings are called ray diagrams.

A ray diagram showing how light moves away from a light bulb

When light waves meet objects, they can either be absorbed, reflected or pass straight through them. We can categorise materials into three groups:

Type of material	Example	What happens to light waves
Transparent	Glass	All or almost all waves pass through (so you can easily see through them) and so none or only a few are absorbed or reflected
Translucent	Tissue paper	Some waves pass through (so you can see through them but not easily) and so some are absorbed or reflected
Opaque	Wood	No waves pass through (so you cannot see through them) and so all are absorbed or reflected

Reflection

A sound wave reflects to form an echo. A light wave reflects in the same way to form a **reflection**. Light waves are reflected from smooth objects like flat mirrors in a very regular way. This is called **specular reflection**. Light waves are reflected from objects without flat surfaces, such as water with faint ripples on it, in an irregular way. This is called **diffuse reflection**.

We can see clear images from surfaces that allow specular reflection

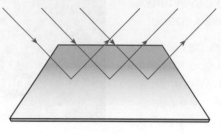

We can only see unclear images from surfaces that allow diffuse reflection

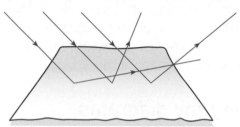

We can use ray diagrams to explain how specular reflection occurs when light is reflected from a mirror. The same diagrams can also show how water waves are reflected.

Both light and water waves have specular reflections when hitting flat surfaces

light waves   water waves

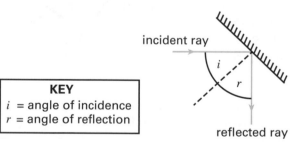
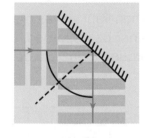

incident ray

KEY
i = angle of incidence
r = angle of reflection

reflected ray

We call the light ray that hits the mirror the **incident ray** and the angle it hits the **angle of incidence**. We call the light ray that reflects the **reflected ray** and this angle the **angle of reflection**. In specular reflection, the angles of incidence and reflection are always the same. This is the **law of reflection**.

Refraction

When light travels from one medium, such as air, into another transparent or translucent medium with a higher density, such as water, it slows down. If it passes from one medium with a higher density, such as glass, into a medium with a lower density, such as air, it speeds up.

On separate cards write *specular reflection, diffuse reflection, refraction,* definitions of each of these key terms and diagrams to represent them. Mix the cards up and then try to match the terms to the diagrams and descriptions.

If this movement is at right angles to the surface then it will change speed but not direction. However, if this movement is at any other angle then the change in speed means a change in direction as well. This is **refraction**.

We see a break in a straw when in a glass of water because of refraction

If light moves from a less to a more dense medium, the ray changes direction towards the normal. If it moves from a more to a less dense medium it changes direction away from the normal.

Make a model of a car travelling from a road into a muddy field to model refraction. Practise going in both directions (mud to road and road to mud). Draw on the normal lines to help you remember whether the light ray moves towards or away from the normal. What happens when both front wheels hit the mud at the same time?

The eye

The eye is your organ of vision. It detects light, allowing you to see.

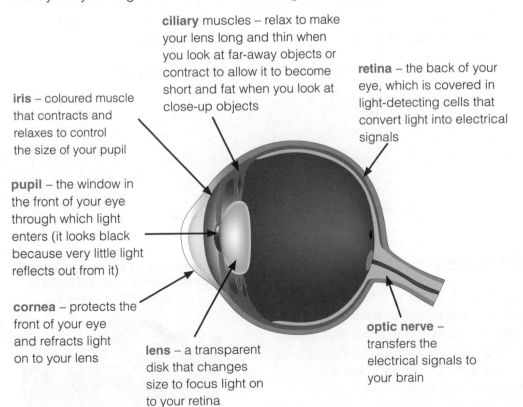

ciliary muscles – relax to make your lens long and thin when you look at far-away objects or contract to allow it to become short and fat when you look at close-up objects

retina – the back of your eye, which is covered in light-detecting cells that convert light into electrical signals

iris – coloured muscle that contracts and relaxes to control the size of your pupil

pupil – the window in the front of your eye through which light enters (it looks black because very little light reflects out from it)

cornea – protects the front of your eye and refracts light on to your lens

lens – a transparent disk that changes size to focus light on to your retina

optic nerve – transfers the electrical signals to your brain

Other animals have different eye structures from ours, such as this compound eye in a fruit fly

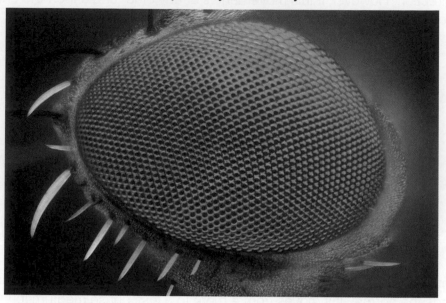

Extend one finger and stretch your arm as far away from you as you can. Focus on this finger with both eyes. Move it slowly towards you until it starts to look blurry. At this point, your lens cannot become any shorter and fatter. Light is not focusing correctly on your retina and all you see is a blur.

Convex lenses

The lenses in your eyes are **convex** in shape. They are curved on both sides, which means that they refract light together into a focal point. This is on your retina in your eye when you see clear images. Convex lenses are used in eyeglasses, magnifying glasses, microscopes and telescopes.

We have two eyes to help us to judge distances. Extend one finger on each of your hands and stretch your arms to each side, as far away from you as you can. Move your arms towards each other until your fingers touch. Now try this with only one eye open. Why is this harder?

How convex lenses like those in your eyes refract light

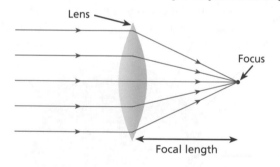

Cameras

A pinhole camera is a simple camera without a lens, which allows light to pass into it through a small hole and makes an upside down image on the back of it. Our eyes (which do have a lens to focus the light) work in a very similar way. Photographic paper is coated with a light-sensitive chemical compound which reacts with the energy in light to turn black to take negatives in film cameras or simple images in pinhole cameras. In this way, it is similar to the light-sensitive cells in your retina.

A pinhole camera allows light to pass through a small hole and focus an upside down image

You can make a pinhole camera by following simple instructions found on the internet. Search for 'How to make a pinhole camera' to find a variety of links.

Dispersion

White light can be split using a prism in a spectrum of colours: red, orange, yellow, green, blue, indigo and violet. The wavelength of these changes from the longest red to the shortest violet. This is **dispersion**. It occurs because red light refracts a little less than orange, which refracts a little less than yellow, and so on.

White light is dispersed into the colours of the spectrum in a prism

Dispersion in individual droplets of rain causes rainbows

Colour

White light is made when the colours in the spectrum come together. It can also be made by mixing the primary colours of light as in the diagram shown here. These are red, green and blue. This is not the same as mixing paint, when you would get a much darker colour.

Any object that is coloured absorbs all the other colours of the spectrum and only reflects that colour. So a red football shirt only reflects red light and absorbs the other colours. White-coloured objects absorb no colours and reflect them all. Black-coloured objects absorb all colours.

Coloured filters only allow the colour of light from which they are made to pass through them. So a red filter absorbs all other colours except red, which it allows to pass through. It 'filters out' the other colours, so we see this footballer's strip as red.

Stick cellophane sweet wrappers over the end of a torch to shine different colours on to a wall (one at a time). You will see that the wrappers are acting as translucent colour filters.

1. State the type of wave that light is.
2. Describe the function of the lens.
3. Explain how you see a green tree.

Progress Check

Worked questions

a)

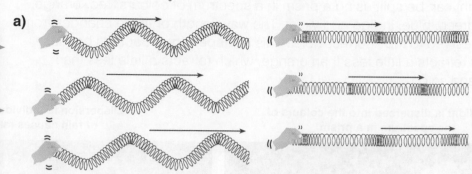

State what type of waves the diagrams show. *(2 marks)*

Transverse and longitudinal.

b) Give an example of each type of wave. *(2 marks)*

Light is transverse. Sound is longitudinal.

c) A wave on the surface of water

Add labels to this diagram to show the positions of amplitude and wavelength. *(2 marks)*

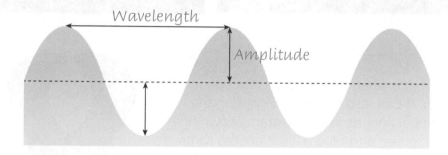

d) Explain how sound and light waves are different. *(4 marks)*

Sound waves cannot travel through space. But light waves can. Sound waves are longitudinal. Light waves are transverse. Both sound and light waves can reflect.

Practice questions

1. This question is about sound waves.

   a) State the range of human hearing. *(1 mark)*

   b) State the units of wavelength and frequency. *(2 marks)*

   c) *(2 marks)*

   These are diagrams of a loudspeaker and a microphone. Describe how the loudspeaker makes sounds and the microphone detects them.

   d) Describe two uses of ultrasound. *(2 marks)*

   e) *(3 marks)*

   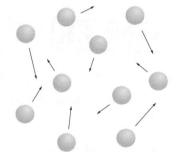

   These are particle diagrams of solids, liquids and gases. Use these diagrams to explain why sound waves travel fastest in solids.

   f) Explain why sounds cannot travel in space. *(3 marks)*

2. This question is about hearing.

   a) *(5 marks)*

   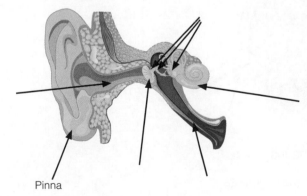

   Pinna

   Label the parts of the ear.

   b) Describe the function of the ear drum. *(1 mark)*

   c) Explain why loud noises can affect your hearing. *(2 marks)*

3. This question is about light waves.

   a) State the definition of a light year. *(1 mark)*

   b) *(3 marks)*

   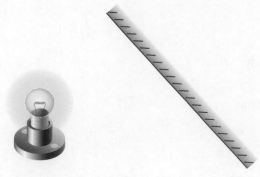

   This ray diagram shows light leaving a bulb. Complete the ray diagrams to show how light reflects from the mirror.

   c) State and describe the two types of reflection. *(2 marks)*

   d) Explain how you know that light waves travel faster than sound. *(1 mark)*

   e) *(5 marks)*

   DANGER

   This is a warning sign with red lettering. Explain how you see both the red lettering and the white background. Explain what the sign would look like if you looked at it through a red filter.

4. This question is about seeing.

   a) Label the parts of the eye. *(5 marks)*

   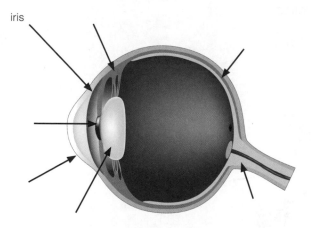

   iris

   b) Describe the function of the iris. *(1 mark)*

   c) Explain why some people wear glasses. *(2 marks)*

After completing this chapter you should be able to:
- describe electric current, voltage and resistance
- explain why static electricity occurs and explain its uses
- describe how magnets are made and explain their uses.

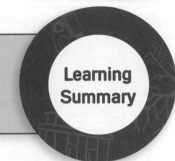

Learning Summary

18.1 Electricity

62

Parts within electrical circuit diagrams are called **components**.

Component	Diagram
Cell	
Battery (two or more cells joined together)	
Open switch	
Closed switch	
Bulb or **lamp**	
Motor	
Ammeter	
Voltmeter	
Resistor	

Make flashcards of different circuit symbols. Draw a symbol on one side of a card and then write the name of the component on the other sides. Test yourself on the names of the symbols using the cards. Keep going through them till you get them all right. Wait a few days and then test yourself again to see if you have can remember the names of all the circuit symbols.

Series and parallel circuits

There are two types of electrical circuit. In a series circuit, all the electricity flows through one path. Parallel circuits have branches to allow the electricity to flow in two or more paths.

Series circuit   **Parallel circuit**

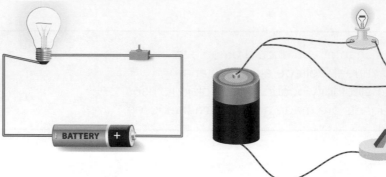

If a break occurs in a series circuit, like turning a switch off or a bulb blowing, the electricity cannot flow and all other components will not work. If this happens in a branch of a parallel circuit, the electricity will still be able to flow in the other branch or branches and any components there may still work.

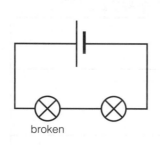

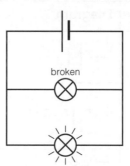

If one bulb breaks in a series circuit all other bulbs will not work, but if one breaks in a parallel circuit the others may remain on: LIGHTS ON

You will learn more about electrons in Topic 8.2.

Current

Electrical **current** is the flow of charge around a circuit. **Electrons** carry electrical energy from the cell or battery. The cell or battery pushes the electrons (and so the current) around the circuit. Current is measured in amperes (or amps) using an ammeter in series. It is the same at all points of a series circuit. Current is not 'used up' in a circuit.

In a parallel circuit, the current is the same before and after the branches. However, it is lower in the wires after they have branched. If the branches have the same components on them then the current reduces equally between them. The sum of the current in the

The current is the same at all points in a series circuit

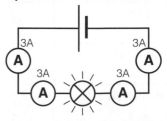

The current is the same before and after the branching on a parallel circuit but reduces on the branches

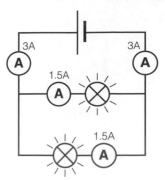

branched wires is the same as before
or after the branching.

Current can flow through conducting materials like metals and graphite where
electrons are free to move. It cannot flow through insulating materials, such as
wood, plastic and rubber, where the electrons are not free to move.

All electrical wires are full of millions of electrons ready to flow when a voltage
is applied by the cell or battery. In the same circuit, thicker wires allow more
electrons to pass through them at one time and so can carry a larger current
than thinner wires.

Potential difference (voltage)

Potential difference tells us how much electrical energy
can be carried around a circuit by the flow of charge
(current). Energy is transferred from the cell or battery,
which pushes the current around to the components
in the circuit. Potential difference is measured in volts
using a voltmeter in parallel across a cell, battery or any
component.

Voltmeters only work across two points in any circuit with
different potential differences. So in the circuit below they
can only be used across the cell or the bulb, because these
are the only two places where there is a potential difference
in electrical energy.

**Electrons leave the cell with electrical energy which
is converted to light energy in the bulb before
returning to the cell to pick up more electrical energy**

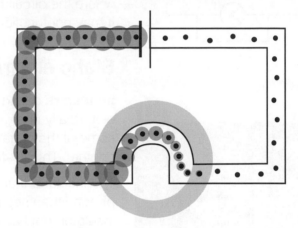

In any circuit with one component, the potential difference is the same across
the cell or battery and this component. In any series circuit with more than
one component, the sum of the potential differences across the components
is the same as across the cell or battery. In a parallel circuit, the potential
difference of the cell or battery is the same as the potential difference across
either or any of the
parallel paths.

**The potential difference is the same across a cell and
components on both branches of a parallel circuit**

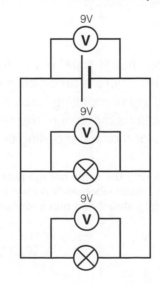

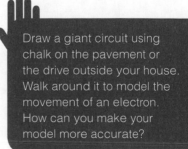

Draw a giant circuit using
chalk on the pavement or
the drive outside your house.
Walk around it to model the
movement of an electron.
How can you make your
model more accurate?

Resistance

Resistance is a measure of how easily the flow of electrons (current) can move through a component. A component with a high resistance slows current more than a component with a lower resistance. This means less energy arrives and so the component does not function as well. This would mean that a bulb is dimmer or a motor is slower.

Resistance is the ratio of potential difference to current. It is difficult to measure so it is usually calculated. Its units are ohms.

In the same circuit, thicker wires can carry a larger current than thinner ones and so have a lower resistance.

Current tends to take the easiest route around a circuit. The easiest route is where the circuit has the least resistance. If this occurs around a component it may not function. This is called a short circuit.

The current takes the path of least resistance and so short circuits the bottom bulb

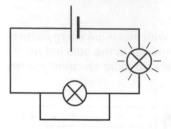

Static electricity

Insulators cannot **conduct** electrical current but they can become electrically charged. If two insulators are rubbed together, friction can remove some of the negatively charged electrons from one and transfer them to another. This is **static electricity**.

The insulator that has had electrons removed will now have more positive protons than negative electrons and so will be positively charged. The insulator that has had electrons added will now have more negative electrons than positive protons and so will be negatively charged.

The friction from rubbing two different insulators with cloths can remove or add electrons building up static electricity

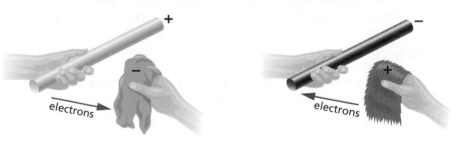

Insulators that have the same charge **repel** each other and those that have opposite charges **attract**. There is an **electric field** between the two insulators in which the attracting or repelling force acts. If insulators with opposite charges touch, electrons move from the negatively charged material to the positively charged one, cancelling out the charges.

Two rods with the same charge repel each other and move apart whereas two rods with different charges attract and move together

Inflate a balloon and rub it with a cloth or against a sofa. The balloon will become charged. If you bring the balloon close to your hair (or someone else's) you will see it attract the hair. You can extend this investigation by tearing up a piece of paper into very small pieces. The charged balloon should attract these pieces and pick them up if you bring the balloon into close enough contact with it. Is there any way you can increase the number of pieces of paper the balloon will pick up?

If you wear an insulating material like a woollen jumper and it rubs against another like some car seats you can build up a static charge. The extra negative charges move from you towards the ground to equalise and gives you an electric shock when they do.

Lightning is formed from the friction between particles of clouds. When sufficient static charge has built up a large amount of electrical energy rushes towards the ground as a spark

We use static electricity in laser printers and electrostatic painting of cars

1. State the symbol for an ammeter.
2. Describe how current varies in a series circuit with one light bulb.
3. Explain the difference between current and voltage.

Progress Check

63

18.2 Magnetism

You will learn more about elements in Topic 8.2.

Iron, nickel and cobalt are the only three **magnetic** elements. Metals that are alloys of these can also be magnetic.

A piece of one of these metals will always be attracted to another **magnet** but it will not necessarily be a magnet. If it is not a magnet, it will still act like one when it is near another magnet and so is called a **temporary magnet**. It can be turned into a **permanent magnet** if it is repeatedly stroked by another magnet. This lines up tiny parts of the metal called **domains** and means that the magnet will now repel other magnets and not just attract them.

Investigate the attraction and repulsion between two magnets and a lump of unmagnetised iron. Imagine that all three looked the same. How could you prove which one was not a magnet?

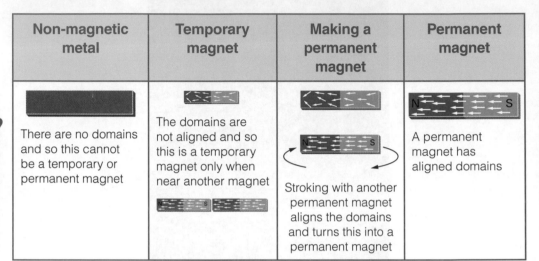

Non-magnetic metal	Temporary magnet	Making a permanent magnet	Permanent magnet
There are no domains and so this cannot be a temporary or permanent magnet	The domains are not aligned and so this is a temporary magnet only when near another magnet	Stroking with another permanent magnet aligns the domains and turns this into a permanent magnet	A permanent magnet has aligned domains

Drawing magnetic fields

Iron filings and plotting compasses line up along magnetic field lines

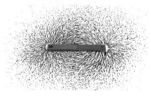

Magnetic fields are the areas around magnets where magnetic metals experience a force. These can be seen by using iron filings or plotting compasses.

The strongest magnetic fields are where field lines are closest. This is at the ends, or **poles**, of magnets. Opposite poles of magnets attract and negative poles repel.

Two of the same poles repel each other and move apart whereas two opposite poles attract and move together

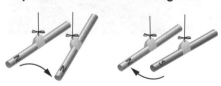

The magnetic field lines between opposite and alike poles

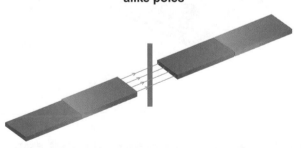

We draw field lines along the path of the field from the north to the south pole. Each line has one arrow on it.

The Earth's magnetic field

The outer core of the Earth is molten iron and nickel which creates a magnetic field. **Compasses** are small pieces of magnetic metal that are free to spin (just like the previous diagrams). These align themselves with the Earth's magnetic field to point towards north.

Understanding the structure of the Earth in Topic 14.1 will help you with this topic.

Electromagnets

An electric current passing through a wire produces a magnetic field.

Plotting compasses show the direction of a magnetic field when a current runs through a wire

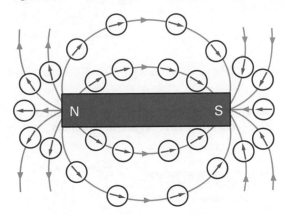

You can make an electromagnet by following simple instructions found on the internet. Search for 'How to make an electromagnet' to find a variety of links.

This field is only produced when the electric current passes through the wire and so we call this an **electromagnet**. These are often coiled to increase the strength of the field. The strength of the magnetic field can be further increased by:
- increasing the current
- adding more coils
- adding an iron core.

Electric bells have an electromagnet which is switched on and off to make the bell ring

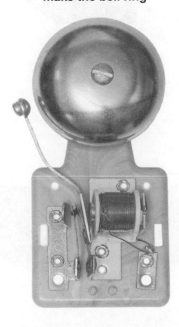

Electromagnets can be switched on and off, making them ideal for use in scrap yards

Motors

If a magnet moves through a coil of wire then a voltage is induced in the wire, creating an electric current. Similarly, if a coil of wire is rotated within a magnetic field then a voltage is also induced and a current flows. This is how simple electric **motors** are made.

Moving the magnet into the coil induces a current in one direction

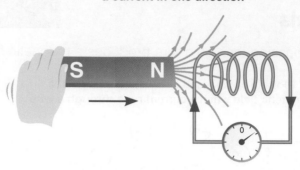

Moving the magnet out of the coil induces a current in the opposite direction

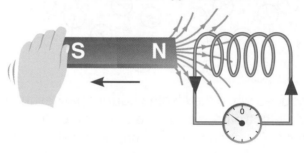

Electric motors are made from rotating coils of wire in a magnetic field

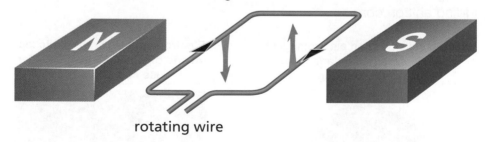

rotating wire

Progress Check

1. State the three magnetic elements.
2. Describe how you can increase the strength of an electromagnet.
3. Explain how compasses work.

Worked questions

a) Describe the difference between series and parallel circuits. *(2 marks)*

Parallel circuits have branches whereas series circuits do not.

b) *(3 marks)*

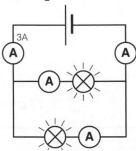

This parallel circuit has identical bulbs on each branch. State what readings the other ammeters would have.

c) State what ammeters and voltmeters measure and describe how they should be used. *(4 marks)*

Ammeters measure current. Voltmeters measure voltage. Ammeters should be used in series and voltmeters in parallel.

d) Explain the only test to determine a permanent magnet. *(3 marks)*

Repulsion is the only test. A magnet is attracted to a lump of iron. Only two permanent magnets repel when like poles are placed close together.

a) Two marks are awarded for this correct answer. It is important to remember that you should not write about the direction of the current. It can flow in both directions around both types of circuit depending upon which way the cell or battery is facing.

b) Three marks are awarded for this correct answer. The current reduced on the branches of this parallel circuit. The current in each branch is 1.5 Amps (because the bulbs are the same on both branches).

c) Two marks are awarded for stating that ammeters measure current and voltmeters measure voltage. It would have been better if the answer include the units (amps and volts) and used the correct term, potential difference, not voltage. Two further marks are awarded for describing that ammeters are used in series and voltmeters in parallel. It would have been better to have said in parallel around a component.

d) One mark is awarded for stating repulsion. However this is not an explanation. A second mark is awarded for explaining that a magnet is attracted to iron (or nickel or cobalt). The final mark is for stating that only two permanent magnets will repel when like (the same) poles are brought together.

Practice questions

1. This question is about electrical circuits.

   a)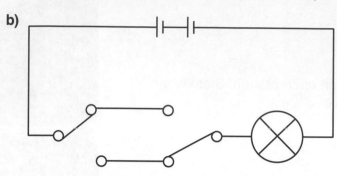

   State the names of these four electrical components. *(4 marks)*

   b)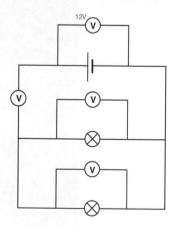

   This circuit uses two switches to control one bulb. Complete the table to show whether the bulb is on or off. The first one has been completed for you. *(3 marks)*

Switch A	Switch B	Bulb
Up	Down	Off
Up	Up	
Down	Up	
Down	Down	

   c) State the definition of potential difference. *(1 mark)*

   d) This parallel circuit has identical bulbs on each branch. State what readings the other voltmeters would have. *(3 marks)*

2. This question is about static electricity.

   a) State how static electricity is formed. *(2 marks)*

   b) Explain what two plastic rods charged with static electricity would do if they were allowed to swing near each other. *(2 marks)*

3. This question is about magnetism.

    a) State what the tiny particles inside magnetic metals are called. *(1 mark)*

    b) Describe how temporary and permanent magnets are made. *(2 marks)*

    c)

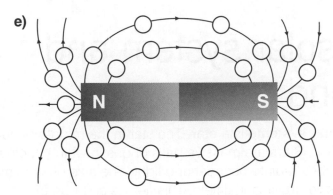

    Draw magnetic field lines to show the magnetic field when the same poles of two bar magnets are facing each other. *(3 marks)*

    d) State where the magnetic field of a bar magnet is strongest and explain how you can tell this. *(2 marks)*

    e)

    Plotting compasses can also be used to show magnetic field lines. Draw the needles on the plotting compasses to show these lines. *(4 marks)*

    f) Iron, nickel and cobalt are the only three magnetic elements. Explain why steel can be magnetised. *(1 mark)*

4. This question is about electromagnets.

    a) State an advantage that electromagnets have over magnets. *(1 mark)*

    b)

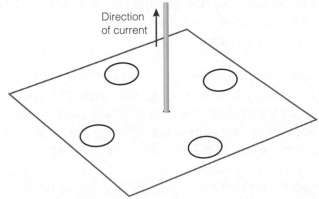

    An electric current passing through a wire produces a magnetic field. Plotting compasses can also show these field lines. Draw the needles on the plotting compasses. *(4 marks)*

    c) Describe the two ways in which a voltage can be induced in a coil of wire. *(2 marks)*

After completing this chapter you should be able to:
- describe the solar system and explain how it fits into our galaxy and the universe
- explain why we have days, years and seasons
- describe weight as the force due to gravity that acts upon a mass.

19.1 The solar system and beyond

64

The Milky Way is a spiral galaxy and our solar system is found in the middle of one of its 'arms'

In the middle of all **solar systems** is a **star**. The star in the middle of our solar system is called the Sun. Stars are massive balls of gas in which nuclear fusion occurs and so they radiate out heat and light. The size of stars means that they have a massive gravitational field. This field is so large that it can keep planets (like Earth), asteroids and comets in orbit around it. Moons (like our own) are objects that are kept in orbit around planets because of their own gravitational field.

Ptolemy (about 90–168CE) was a Greek astronomer who created a model of the solar system called the geocentric model. Earth was in the middle of this model and the Sun and other planets moved around it. Many years later, Copernicus (1473–1543) used telescopes (which had not been invented in Ptolemy's lifetime) to come up with another model. This was called the heliocentric model and has the Sun in the centre of the solar system. This is an example of how new evidence has allowed scientists to develop their theories.

Radiation is covered in Topic 15.6 and reflection in Topics 17.2 and 17.3.

Stars are the only objects in space that give out heat and light. Objects which do this are called **luminous**. Planets, moons, asteroids and comets all shine in the night sky because they reflect the light from the stars. Objects that don't give out light are called **non-luminous**. Without the heat and light that radiates from our Sun, there would be almost no life on Earth.

Pluto was discovered in 1930 and was designated as the ninth planet from the Sun. In 2006, the International Astrological Union formally defined planets as: (a) being in orbit around the Sun, (b) having sufficient mass to be nearly a round shape, and (c) having sufficient gravity to pull objects near it into it. Pluto did not meet this final criterion and so it is now called a dwarf planet.

Our Solar System

Mercury | Venus | Earth | Mars | Jupiter | Saturn | Uranus | Neptune

Lots of galaxies make up the universe

	Relative size
The Sun is the star at the centre of our solar system. It is so big it makes up more than 98% of all the mass in our solar system	110
Mercury is a dense, rocky planet with lots of craters. It has no atmosphere and so gets very hot in the day and cold at night	0.4
Venus is an extremely hot and dry, rocky planet with a dense carbon dioxide atmosphere. It has a pressure nearly 100 times that of Earth	0.9
We have life on **Earth** because water exists as a liquid. We are too close for all the water to be frozen and too far for it all to be steam	1
The **Moon** is a tiny ball of rock that orbits the Earth. It has no atmosphere	0.25
Mars has lots of iron oxide (rust) on its surface and so it looks red. It has polar ice caps like Earth but the rest of it is covered in dry, red dust and rocks	0.5
Jupiter has no solid surface and is made of hydrogen and helium gas. It has powerful storms on its surface which last for hundreds of years	11
Saturn also has no solid surface and is made of hydrogen and helium. It's moon, Titan, is the only one in our solar system with an atmosphere	9.4
Uranus also has no solid surface and is made of hydrogen and helium. It has an atmosphere of frozen water, ammonia and methane	4
Neptune also has no solid surface and is made of hydrogen and helium. It has an atmosphere of frozen water, ammonia and methane	3.8

Millions of solar systems, each with their own star at their centre, are grouped into giant structures called **galaxies**. The stars that you see in the night sky are in the middle of other solar systems within our galaxy, which is called the Milky Way. Telescopes can show us other galaxies. They often look like small, blurry stars. All of the galaxies that exist make up the **universe**.

Using paper and string construct a hanging mobile showing the Sun, Earth and Moon and their relative positions. You can extend this activity by adding in the other planets in the Solar System.

1. State the name of the star in the middle of our solar system.
2. Describe what is beyond our solar system.

Progress Check

65

19.2 Days, years and seasons

Days and years

A day on Earth is 24 hours. This is the time that it takes for the Earth to rotate once on its own axis. Other planets in our solar system rotate at different speeds and so have different day lengths. The planets that rotate quickest are Jupiter and Saturn, on which a day is only ten hours. Venus is the slowest rotating planet and one day on Venus takes nearly 6000 hours.

When the Sun faces the part of Earth that you are on, its light shines on you and so it is your daytime. When the part of Earth that you are on has rotated around completely to face away from the Sun, no light shines on you and this is night time. So day and night are caused by the Earth's rotation on its axis.

A year on Earth is 365.25 days. This is the time that it takes for the Earth to rotate once around the Sun. This time depends upon how fast a planet moves and how far from the Sun it is. The further from the Sun the bigger the orbit it has. Mercury has the quickest year, at 88 Earth days, whereas it takes 165 Earth years for one Neptune year (or for Neptune to rotate once around the Sun).

> Can you calculate how old you would be in years if you had been born on Mercury and not on Earth?

Seasons

The Earth is tilted at 23.5 degrees and this tilt causes our seasons. The summer for the northern hemisphere is between June and August. At this time, the northern hemisphere is tilted towards the Sun. This means that more time is spent in daylight and more of the Sun's rays are concentrated on the surface, making it hotter. In these months, the southern hemisphere is tilted away from the Sun and so less time is spent in daylight and fewer of the Sun's rays are concentrated on the surface. The opposite is also true for both hemispheres. The equator points towards the Sun for most of the year and so seasons are less obvious.

> Use a football and a torch to model the Sun's rays falling on the equator and then on the northern and southern hemispheres. Can you explain this to your parents or friends?

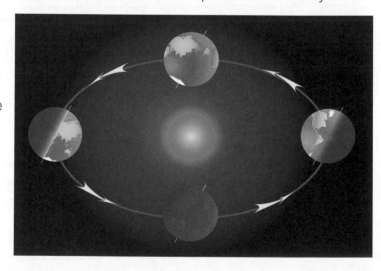

Progress Check

1. Describe what a day and a year are.
2. Explain why we have seasons.

19.3 Gravity and its effects

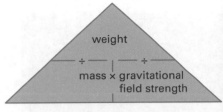

Gravity is the force of attraction between any two objects. Anywhere that this force can be felt is within the gravitational field of the object. The bigger the object the larger the gravitational force. Objects have to be huge, like moons or planets, for us to notice their gravitational effect. Gravity keeps:

- us attracted to the surface of the Earth
- the Moon orbiting around the Earth
- the Earth orbiting around the Sun.

Weight is name given to the force that acts upon the **mass** of an object. So your weight is the force of the Earth's gravitational field pulling your mass downwards. We often say that our weight is measured in stones but it is really measured in newtons, like all other forces.

> weight (N) = mass (kg) × gravitational field strength (N/kg)

You can investigate the effects of gravity and air resistance on masses by attaching homemade parachutes (made from plastic bags) to objects in your house. As you do this, think about what would happen to your parachute if you dropped the same object on other planets. Think carefully about what slows a parachute down. (Why might it not work on Mercury?)

This bag of sugar has a mass of one kilogram and so a weight of ten newtons on Earth

The Moon has a gravitational field strength of 1.7 and so the same bag of sugar would have a weight of 1.7 newtons on the Moon

On Earth, the gravitational field strength is 10 N/kg. On bigger planets in our solar system it is larger and so your weight would be heavier as well. On smaller planets and our Moon the gravitational field strength is smaller and so your weight would be lighter. This is why astronauts seem to jump when they walk on the Moon. Mercury has the smallest gravitational field of the planets in our solar system. You would lose weight simply by going there.

Mercury

1. State the scientific units of weight.
2. Explain why astronauts bounce on the Moon.

Progress Check

Worked questions

a)

pathway of Sun
in winter

east

west

a) One mark is awarded for drawing the pathway of the Sun in summer higher than in winter. One mark is awarded for beginning and ending this pathway at the same places in summer as in winter. A final mark is awarded for showing that the pathway of the Sun moves from east to west.

The diagram above shows the pathway that the Sun takes in winter. Draw on a line to show how it moves in summer. Draw an arrow to show the direction of the movement. *(3 marks)*

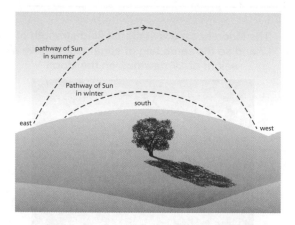

b) The answer is awarded one mark for explaining the Earth rotates on its axis. A day is one full rotation. You must be careful not the confuse years with days here. A year is the time it takes the Earth to orbit the Sun once.

b) Explain why the Sun moves across the sky. *(1 mark)*

The Earth rotates on its axis and this is what makes it move across the sky during the day.

c) One mark is awarded for explaining that shadows are areas where light cannot reach because it is blocked by an object. A second mark is awarded for explaining that shadows are shorter in summer because the Sun is higher in the sky.

c) Explain why shadows are formed and whether they would be longer in summer or winter. *(2 marks)*

Shadows are formed when sunlight is blocked by an object. They would be shorter in summer because the Sun is higher in the sky.

d) One mark is awarded for stating that the closest satellite experiences the greatest force of gravitational attraction. One mark is also awarded for explaining that satellite A has the smallest orbit because it is closest. One additional mark would have been awarded for explaining that it would have the quickest orbit time because it moved through space at the same speed but had less distance to cover.

d)

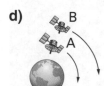

This diagram shows two satellites orbiting the Earth at the same speed as the Moon. Explain which one experiences a greater gravitational force. Explain which one orbits in the shortest time. *(3 marks)*

Satellite A experiences a greater force because it is closer to the Earth. Satellite A has the smallest orbit because it is closest. Satellite A moves at the same speed as B but has a shorter distance to travel so travels in one orbit in a shorter time.

Practice questions

1. This question is about space.

    a)

    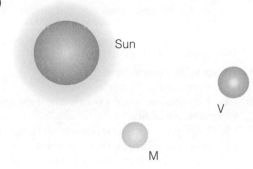

    State the names of planets M and V.   *(2 marks)*

    b) State whether each of the five objects in the diagram is luminous or
       non-luminous.   *(5 marks)*

    c) The sizes of the objects in the diagram above are not drawn to scale. Describe what else
       is misleading about this diagram.   *(1 mark)*

    d) Shade one side of the picture of Earth to show it in night time.   *(1 mark)*

    e) Explain the similarities and differences between the four planets closest to the Sun
       and the four furthest from the Sun.   *(4 marks)*

2. This question is about gravity.

    a) State the equation that is used to calculate weight. Include the units of all terms. *(3 marks)*

    b) The Sun is approximately 110 times bigger than the Earth and makes up 98%
       of the mass of the solar system. The Moon is a quarter of the size of the Earth
       and has a much smaller mass. Use the equation above to describe how your
       weight would be on the surface of the Sun and the Moon.   *(4 marks)*

    c) Explain why a hammer and a feather dropped by an astronaut would hit the
       surface of the Moon at the same time.   *(3 marks)*

Answers to progress check questions

1.1 Plant and animal cells
1. Cell wall, vacuole and chloroplast.
2. Walls provide structure for plant cells and membranes control what enters and exits all cells.
3. They have no nucleus and a biconcave shape to increase the surface area to absorb as much oxygen as possible.

1.2 How animal and plant cells make up organisms
1. Cell, tissues, organs, organ systems and then organisms.
2. An organ is a group of tissues in the same place that complete the same function.

1.3 Unicellular organisms
1. An organism that is one cell is size, e.g. *Streptococcus* or yeast.
2. Plasmid DNA, cell wall and flagella.

2.1 The skeleton and muscular systems
1. In the marrow of your large bones.
2. One bone has a ball on one end and the other a socket allowing rotation, e.g. hips or shoulders.
3. Tendons connect bones to muscles. When muscles contract they pull on tendons to move bones like levers.

2.2 The digestive system
1. In your mouth.
2. Your small intestine absorbs food and your large intestine absorbs water into your blood.
3. Rings of muscle contract behind food to propel it through the digestive system.

2.3 The gas exchange system
1. Diffusion.
2. Your intercostal muscles contract to move your ribcage up and out. Your diaphragm contracts to move it downwards. This increases your chest capacity and so air rushes in to your lungs.
3. It has tough rings of cartilage surrounding it which keep it open at all times.

3.1 A healthy and balanced diet
1. Carbohydrate.
2. Increased weight leading to obesity and related health problems, e.g. type 2 diabetes and heart disease.
3. It provides a solid substance for our muscles to push along our digestive system during peristalsis

3.2 A healthy gas exchange system
1. Lungs, mouth and throat.
2. It reduces your resting heart rate and quickens your recovery rate after exercise.

3.3 Legal and illegal drugs
1. Stimulants.
2. Almost always this leads to negative consequences for your health or society, such as antisocial behaviour.
3. They depress it, making your reactions slower. This can make activities like driving dangerous.

4.1 The male and female reproductive systems
1. To connect the ovaries with the uterus.
2. To keep them cooler than the body which maximises sperm production.

4.2 The menstrual cycle
1. By the hormones oestrogen and progesterone.
2. The blood vessels lining the uterus thicken in preparation for a fertilised ovum to embed and develop into a baby.

4.3 From fertilisation to birth
1. Sperm and ova (eggs).
2. It is contained within amniotic fluid in the amnion within the uterus.
3. The toxins in cigarettes, alcohol and drugs can pass to the baby. They can slow growth rates, make birth happen to early and even kill unborn babies.

4.4 Sexual reproduction in other animals
1. Birds and reptiles.
2. They release them in water and so they are more likely to be lost.

4.5 Reproduction in plants
1. Plants can reproduce asexually and sexually.
2. It first touches the stigma where a pollen tube is formed. This transports the pollen nucleus down the style to the ovary.
3. They have brightly coloured flowers and/or strong smelling nectar to attract insects for pollination.

5.1 Photosynthesis and plant structures
1.

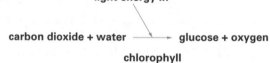

light energy in

carbon dioxide + water $\longrightarrow$ glucose + oxygen

chlorophyll

2. They have root hair cells which increase the surface area of the plant in contact with the soil.
3. Transpiration is the evaporation of water from plant leaves which pulls more water up from the roots. This provides the leaves with water for photosynthesis.

5.2 Aerobic and anaerobic respiration
1.

energy out

glucose + oxygen $\longrightarrow$ carbon dioxide + water

$C_6H_{12}O_6 + 6O_2 \longrightarrow 6CO_2 + 6H_2O$

2. Your muscles would feel 'rubbery' and you may get painful cramp.
3. Only some microorganisms can ferment sugars. This releases energy for them and makes ethanol and carbon dioxide as by-products. Respiration occurs in all other living cells. It also releases energy but makes water and carbon dioxide as by-products.

6.1 Relationships within ecosystems
1. Photosynthesis.
2. Some pesticides cannot be excreted and so concentrate at higher trophic levels.
3. It means that all living things in an ecosystem depend upon each other for their survival.

6.2 Pollution
1. CFCs.
2. Trees and other plants are killed. Statues and other stone structures are eroded.

3. Carbon dioxide and other greenhouse gases build up in the atmosphere. These trap more of the Sun's energy. This is the greenhouse effect and it leads to global warming.

7.1 Heredity and genetic diversity
1. Asexual reproduction.
2. Two different sperm fertilise two different ova, which grow into two genetically different babies.
3. Without it, organisms cannot adapt to changing conditions or disease and so are more likely to become extinct.

7.2 From DNA to cells
1. A section of DNA that is responsible for producing a characteristic.
2. It has a double helix shape and is made from the base pairs A-T, T-A, C-G and G-C.
3. Her X-ray images were crucial in Watson and Crick's discovery but she died before she could receive the Nobel prize.

7.3 Variation
1. Genetic and environmental.
2. Continuous variation produces data in a range whereas discontinuous variation produces data in discrete groups.

7.4 Natural selection and evolution
1. Variation.
2. Some organisms in a population will possess characteristics that mean they are better adapted to survive and reproduce than others. They are more likely to survive.
3. Darwin correctly predicted that the Church would criticise him and his theory because it disagreed with the Bible.

8.1 Solids, liquids and gases
1. The smell particles diffuse across the room.
2. Higher. Increasing temperature increases the pressure of a gas.
3. It will melt into liquid copper and then eventually boil into gaseous copper.
4. To overcome the attraction between the particles.

8.2 Atoms and molecules; elements and compounds
1. S, Mg, Ne, N, K, Al
2. Two atoms of carbon and six atoms of hydrogen.
3. It is not an element.

8.3 Chemical reactions
1. a and c are chemical reactions; b and d are physical changes.
2. Because atoms cannot be created or destroyed in chemical reactions or physical changes.

9.1 Pure substances and mixtures
1. A.
2. Molecules.

9.2 Dissolving
1. Water is the solvent and sugar is the solute.
2. They break up into particles too small to be seen.

9.3 Separating mixtures
1. Filtration.
2. Evaporation.
3. Fractional distillation.

9.4 Testing for purity
No, this sample was not pure.
It suggests that this sample of water was pure.

10.1 Chemical equations
1. Hydrogen peroxide → water + oxygen
2. $H_2O_2 \rightarrow H_2O + O_2$
3. $2H_2O_2 \rightarrow 2H_2O + O_2$

10.2 Energy in chemical reactions
1. Exothermic.
2. Exothermic.
3. Calcium oxide (CaO) and carbon dioxide (CO_2)

10.3 Oxidation and reduction
1. Oxidation.
2. The carbon monoxide (CO) has been oxidised; the iron oxide (Fe_2O_3) has been reduced.

10.4 The speed of reactions and catalysts
1. Substance C.
2. Substances B and D (not significantly different from the control).
3. So that there was something to compare the other substances with/so you know how fast the reaction is without a catalyst.

11.1 Acids, alkalis and the pH scale
1. Acidic.
2. 7.
3. 8, 9, 10 or 11.

11.2 Reactions of acids
1. Nitric acid + lithium hydroxide → lithium nitrate + water
2. Hydrochloric acid + aluminium → aluminium chloride + hydrogen
3. Sulfuric acid + iron oxide → iron sulfate + water
4. Nitric acid + nickel carbonate → nickel nitrate + water + carbon dioxide

11.3 Producing a salt
1. Sulfuric acid and either iron hydroxide, iron, iron oxide or iron carbonate.
2. Because hydrogen is produced when an acid reacts with a metal and carbon dioxide is produced when an acid reacts with a carbonate. No gas is made when an acid reacts with an alkali or metal oxide.

12.1 Properties and uses
1. It is a good conductor of electricity. (It also has a low density and it is ductile.)
2. Non-metal.
3. Hydrogen would be too dangerous (flammable).

12.2 Developing and using the periodic table
1. Bromine
2. Group 3
3. Very violently/quicker than rubidium/explodes.

13.1 Constructing a reactivity series
1. At the bottom.
2. It would be too dangerous.

13.2 Displacement reactions
1. Potassium.
2. Nothing would happen because copper is less reactive than calcium.

13.3 Extracting metals
1. Electrolysis.
2. Displacement using carbon.

14.1 Structure of the Earth
1. The crust.
2. The crust.
3. The outer core.

Answers to progress check questions

14.2 The Earth's crust
1. Aluminium.
2. Platinum (also rhodium and palladium).
3. Cars/aeroplanes/sports equipment.

14.3 The rock cycle
1. From lava on the surface of the Earth/outside a volcano/cooled quickly/extrusive rock.
2. Sedimentary rock.
3. No animals could survive living inside molten rock/any fossil present in sedimentary rock would be destroyed when it was melted to form igneous rock.
4. High temperature and pressure.

14.4 Composition of the atmosphere
1. Carbon dioxide and methane (others include water vapour, nitrous oxide, ozone, CFCs).
2. Nitrogen.
3. Carbon dioxide
4. Coal, oil and natural gas.

15.1 Energy stores and transfers
1. Chemical store in food → kinetic store in moving body + thermal store in body and surroundings.
2. Gravitational store in person on diving board → kinetic store in falling person.
3. Elastic store in stretched rubber band → kinetic store in moving band + thermal store in band and surroundings.
4. Energy is slowly transferred from the pendulum's gravitational and kinetic stores to the thermal store of the surroundings.

15.2 Energy resources
1. Any three from: Sun/biomass/hydroelectric/wind/geothermal/wave/tidal.
2. Coal, oil, natural gas.

15.3 Energy from food
1. Chocolate.
2. Fat.

15.4 Power and appliances
1. 100 W bulb.
2. 18 p.

15.5 Machines and work
1. $20 \times 0.1 = 2$ joules.
2. $2 \div 0.01 = 200$ N.

15.6 Temperature and heat energy
1. Conduction.
2. Convection.
3. White.

16.1 Speed, distance and time
1. 25 m/s.
2. 300 m.
3. 60 s.
4. 5 m/s.

16.2 Relative motion
1. 30 m/s.
2. 5 m/s.

16.3 Forces
1. 650 N, acting downwards.
2. 800 N, acting upwards.
3. 0 N.
4. When she reaches the ground, because the resultant force is greater.

16.4 Hooke's law
1. 15 cm.
2. Because when there is no force applied, there is no extension of the spring.

16.5 Moments
1. Your sister must move closer to the pivot.
2. 6000 Nm.

16.6 Non-contact forces
1. Gravity, magnetism, static electricity.
2. The mass of the two objects and the distance between them.

16.7 Pressure in fluids
1. The tip of the pin has a smaller area, so the pressure on the board is much greater than on your finger.
2. As the air pressure outside the balloon decreases, the balloon expands.

17.1 An introduction to waves
1. Longitudinal and transverse.
2. Longitudinal waves are started by a movement in the direction of the wave, whereas transverse waves are started by a movement at right angles to it.
3. Sounds are vibrating waves of particles and so require a medium to travel in. The vacuum of space has no particles to vibrate.

17.2 Sound waves
1. Amplitude.
2. To collect sound vibrations from the surrounding air.
3. Sounds are waves of vibrating particles. Solids have particles closer together than liquids or gases and so sounds travel fastest through them.

17.3 Light waves
1. Transverse.
2. A transparent disk that changes size to focus light on to your retina.
3. A green tree absorbs red, orange, yellow, blue, indigo and violet light and only reflects green light into your eye.

18.1 Electricity
1.

2. The current remains the same throughout a series circuit.
3. Current is the flow of charge through a conductor. Voltage is a measure of how much energy the current is carrying.

18.2 Magnetism
1. Nickel, iron and cobalt.
2. Increase the current, add more coils or include an iron core.
3. Compasses are magnets that are free to rotate. They align themselves with the Earth's magnetic field and so point north.

19.1 The solar system and beyond
1. The Sun.
2. Many other solar systems make up our galaxy, the Milky Way. Many other galaxies beyond this make up the universe.

19.2 Days, years and seasons
1. A day is the time that it takes for the Earth to rotate once on its axis and a year is the time it that takes the Earth to rotate around the Sun.
2. We have seasons because the Earth is tilted. This means that we have hotter periods in the year where we are closer to the Sun and colder ones when we are further away.

19.3 Gravity and its effects
1. Newtons.
2. The Moon is has less mass than the Earth and so has less gravity.

Chapter 1

1. a) Root hair cell.
   b) It has a long hair, which increases the surface area.
      This allows more water to be absorbed.
   c) Ciliated cell.
   d) It has tiny hair-like structures (cilia) which wave to remove dirt and bacteria.
      This keeps our gas exchange system clean.
   e) Sperm cells swim to meet the ovum whereas the ovum just floats down to meet the sperm.
      So the sperm cell requires more energy.
      It gets this from respiration in mitochondria. ⊃ 20

2. a) Microscope.
   b) Nucleus / Cytoplasm / Membrane / Mitochondria
   c)

Component	In animal cell	In plant cell
Nucleus	Yes	Yes
Cytoplasm	Yes	Yes
Membrane	Yes	Yes
Mitochondria	Yes	Yes
Wall	No	Yes
Vacuole	No	Yes
Chloroplasts	No	Yes

   d) The membrane controls what enters and exits the cell.
   e) Plant cells require walls for structure.
      Animals have bones or other structures that provide support. ⊃ 18

3. a) The cell.
   b) The heart.
      Muscle tissue and nerve tissue.
   c) Organs are groups of different tissues in the same place that complete the same function.
      Organ systems are groups of organs that work together to complete the same function.
   d) Any unicellular organisms, e.g. yeast or *Streptococcus* ⊃ 21

4. a) Chromosomal DNA / Plasmid DNA / Cell wall / Plasma membrane / Cytoplasm / Flagella
   b) Cellular reactions like respiration occur here.
   c) Some unicellular organisms cause diseases like *Streptococcus*.
      Others like yeast help us make things like bread and alcohol. ⊃ 22

Chapter 2

1. a) Humerus / Ribs / Femur / Fibula
   b) They allow the skeleton to move.
   c) A hinge joint acts like a hinge on a door.
      The neck.
   d) Label pointing to fixed joints in the skull.
   e) The skull protects the brain.
      The rib cage protects the heart, lungs etc. ⊃ 27

2. a) Contract and relax. (No mark to be awarded for push and pull.)
   b) Biceps / Triceps
   c) Bones are connected to muscles by tendons.
      When muscles contract they pull on tendons.
      Muscles work in pairs / Muscle antagonism.
      Tendons then pull on bones which results in movement.

   d) Involuntary.
      To respond quickly and automatically to change (in light levels for your iris).
   e) Antagonistic muscles.
      Muscles can only contract and relax.
      They cannot push backwards.
      So they need to work in pairs.
      When one contracts the second relaxes.
      When the first relaxes the second contracts. ⊃ 28

3. a) Mouth / Oesophagus / Stomach / Liver / Pancreas / Small intestine / Rectum
   b) Food mixes with acid to kill microorganisms.
      Food mixes with enzymes for digestion.
   c) Enzymes are biological catalysts, which means that they speed up reactions.
      Carbohydrase enzymes break down carbohydrates.
      Into sugars.
   d) Peristalsis is the regular contraction of muscles in the digestive system.
      This happens behind food to push it forwards.
      (Award marks if diagram shows contraction of muscle and movement of food.)
   e) Digested food is at a higher concentration in the small intestine than in the blood.
      It moves from this high concentration to a lower one.
      This is called diffusion. ⊃ 30

4. a) Bronchi / Bronchiole / Alveoli / Diaphragm / Rib
   b) Arrows to show:
      • Oxygen moving from the alveolus to the blood.
      • Carbon dioxide moving from the blood to the alveolus.
      • Water moving from the blood to the alveolus.
   c) The substances move from a higher concentration to a lower one.
      This is called diffusion. ⊃ 33

Chapter 3

1. a) Two from: increases the strength of muscles; reduces the risk of heart disease; reduces your resting heart rate and quickens your recovery rate; improves flexibility and strength.
   b) Inflammation of the airways making it hard to breathe.
      Using inhalers or bronchodilators.
   c) It contains the correct proportions of the six food groups and water.
      Obesity and related health problems.
      Weight loss, malnutrition and starvation.
   d)

   e) Smoking breaks down the walls of the alveoli.
      This reduces the surface area of the lungs.
      This makes it harder to get oxygen into the blood.
      This means cells can respire less.
      So regular smokers often have less energy. ⊃ 42

2. a) Substances taken as medicine, to intoxicate or enhance performance.
   b) Stimulants.

c) They speed up the nervous system and so make your reactions faster.

d) Caffeine is legal and cocaine is not.

e) Solvents arrows and labels to: lungs, liver, brain or kidneys.
Alcohol arrows and labels to: eyes, brain, liver or kidneys. ⤺44

Chapter 4

1. **a)** Testes / Sperm duct / Urethra / Foreskin / Scrotum

b) To carry urine from the body.
To ejaculate sperm into the vagina of a women during sexual intercourse.

c) It connects the testes to the urethra.

d) Only one sperm can fertilise an ovum.
It is a difficult journey for sperm to swim to find the ovum. Many die before they reach it.

e) Sperm have to swim to reach the ovum.
Ova float down the fallopian tube to the uterus.
Sperm need more energy from respiration.
Respiration occurs in the mitochondria. ⤺48

2. **a)** Zygote. (Do not accept fetus; this is a developing embryo.)

b) Powerful contractions of the uterus push the baby.
The baby is pushed through the cervix and vagina.

c) The jam jar represents the amnion.
The water represents the amniotic fluid.
The egg represents the fetus.

d) The placenta.
This organ allows food and oxygen from the mother's blood to diffuse into the baby.
This then passes along the umbilical cord. ⤺31

3. **a)** Amphibians / Fish

b) Fertilisation is internal.
The female lays eggs in a nest.
The eggs hatch into chicks.
The parents feed the chicks until they can fly the nest.

c) They produce milk. (Do not accept they care for their young because other animals, including birds, also do this.)

d) They lay them on land, so they are less likely to be lost or not fertilised. A high level of parental care is given. ⤺54

4. **a)** Style / Ovary / Anther

b) By bees or other insects.
By wind.

c) A pollen tube is formed.
This grows through the style to the ovary.
The pollen nucleus moves down this tube.
To fertilise the ovum and form a seed.

d) Two from: wind, animals, water, ejection.

e) Bees are important pollinators of many crops.
Without them our crops will not produce as many fruits or seeds. ⤺55

Chapter 5

1. **a)** Chloroplasts. (Do not accept chlorophyll as the question asks for the component.)

b) $6CO_2 + 6H_2O \rightarrow C_6H_{12}O_6 + 6O_2$
Reactants: carbon dioxide and water
Products: oxygen and glucose

c) Used for respiration.
Formation of cellulose.
Stored as starch.
Formation of proteins.

d) It increases with higher light intensities.
It increases with more water.

It increases with higher carbon dioxide levels.
It increases at the correct temperature.

e) In summer in the northern hemisphere leaves are present on many trees.
In winter, deciduous trees drop their leaves.
There is more photosynthesis in summer.
There is less photosynthesis in winter which means that less carbon dioxide is absorbed. ⤺61

2. **a)** Nucleus / Cytoplasm / Membrane / Mitochondria / Cell wall / Vacuole

b) To provide support for the cell.
Cellulose.

c)

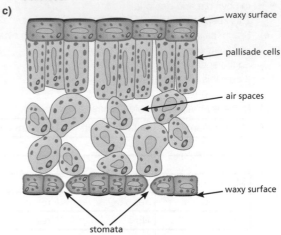

waxy surface
pallisade cells
air spaces
waxy surface
stomata

d) They have root hair cells.
This maximises the contact between the roots and the soil.
Water moves from the soil to the plant.

e) Water evaporates from the leaves of plants.
This pulls water up from the roots. ⤺62

3. **a)** Mitochondria.

b)

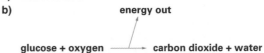

energy out

glucose + oxygen ⟶ carbon dioxide + water

Reactants: oxygen and glucose
Products: carbon dioxide and water

c) Three from: movement, reproduction, sensitivity, growth, respiration, excretion, nutrition, maintaining constant body temperature.

d) They are usually below the ground so do not come into contact with sunlight.
They cannot photosynthesise.
So do not have chloroplasts.
So are white.

e) During the day the levels of oxygen in the air increase.
During the night these levels decrease.
(Award no marks for an explanation for this question asked for a description.)

f) During the day plants and animals are respiring.
But plants are also photosynthesising.
This means that the levels of oxygen increase.
At night plants and animals are also respiring.
But plants are not photosynthesising.
So the levels of oxygen decrease. ⤺65

4. **a)**

energy out

glucose ⟶ ethanol + carbon dioxide

Reactant: Glucose
Products: Ethanol and carbon dioxide

b)

five percent energy out

glucose ————→ lactic acid

Reactant: Glucose
Product: Lactic acid

c) They have used most of the oxygen in their body during respiration.

This respiration created the extra energy needed for their exercise.

They need to breathe deeply to replace this extra oxygen or remove lactic acid.

d) If they run out of oxygen anaerobic respiration will occur.

This creates lactic acid.

Lactic acid builds up and leads to cramp.

e) During the 10 minutes after exercise the person replaces the missing oxygen.

This reacts with lactic acid to make carbon dioxide and water.

This releases the extra energy stored in the lactic acid. ⤶ 66

Chapter 6

1. **a)** The transfer of energy.

b) Energy.

The seven life process: movement, reproduction, sensitivity, growth, respiration, excretion, nutrition.

c) The number of rabbits will increase because they have no predators.

This will mean that the amount of grass will decrease.

There will be less food for the insects and slugs.

There will be fewer insects and slugs so fewer thrushes and voles and fewer hawks as well.

d) Producer: grass.

Primary consumer: rabbit, insect or slug.

Carnivore: fox, voles or thrush.

e) Hydrothermal volcanic vents found on the ocean floor.

Bacteria feed on the chemicals released from the vents.

No light can reach the ocean floor so no plant can survive here. ⤶ 71

2. **a)** Two from: oil, sewage, pesticides, fertilisers, metals.

b) Two from: carbon dioxide, carbon monoxide, CFCs, nitrogen oxides, sulfur dioxides.

c) Carbon dioxide.

This gas is released when fossil fuels are burned.

The amount of fossil fuels being burnt has increased in recent years.

So has the concentration in the atmosphere.

d) Carbon dioxide builds up in the atmosphere.

This allows the Sun's rays into the atmosphere but traps them here for longer.

This is called the greenhouse effect.

This leads to the increasing temperature of the Earth (global warming).

e) During pond dipping you are likely to catch some aquatic organisms.

Some of these are bioindicators.

They tell us if water is polluted.

Bloodworms and sludgeworms are present in polluted water.

Stonefly nymphs are present in clean water. ⤶ 75

Chapter 7

1. **a)** DNA base pairs / Genes / Chromosomes / Genome

b) An entire copy of all your DNA (genes, or chromosomes).

c)

Key
A
T
G
C

d) James Watson and Francis Crick discovered the structure of DNA.

They used X-ray images.

These images were taken by Rosalind Franklin and Maurice Wilkins. ⤶ 81

2. **a)** The differences within one species.

b) Genetic: eye colour.

Environmental: scars.

Genetic and environmental combined: body weight.

c) Discontinuous variation.

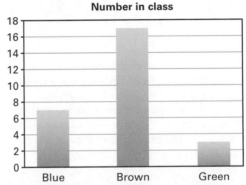

d) More than half of the students had brown eyes.

The smallest number of students had green eyes.

The number of students that had blue was in between those with blue and brown eyes.

e) In a line graph or histogram.

Height is a continuous variable. ⤶ 83

3. **a)** Charles Darwin and Alfred Russel Wallace

b) Darwin visited the Galapagos Islands.

Here he saw populations of different but similar birds on the different islands.

He thought that they came from the same original population from the mainland.

But had slowly changed to adapt to their own islands.

c) Charles Darwin was worried about the reaction of the Church.

d) They did not have advantageous characteristics. They could not evolve fast enough to changing conditions.

e) Originally, the black moths were at a disadvantage on the pale trees.

The dark moths were easier for birds to eat.

So most moths were pale.

Then the trees were turned black because of the pollution.

So the light moths were now easier for birds to eat.

So slowly the black moths became more common. ⤶ 85

4. **a)** Pollen and ova.

b) By replicating their DNA and splitting in two (binary fission).

c) They fuse during fertilisation to form one cell.
Half of this cell's DNA has come from each parent.

d) Your parent's gametes are not identical.
If they were you and your siblings would be genetically identical.
Or you have changes resulting from environmental variation like scarring.

e) Identical twins result from the splitting of one fertilised ovum.
So they must be the same sex.
Non-identical twins result from two different fertilised ova.
So they can have different sexes. ↄ 79

Chapter 8

1. a) Particles have a regular arrangement.
Particles are touching their neighbours.
Particles are vibrating.

b) Melting.

c) The ice absorbs energy.
This energy is used to break/overcome the attractive forces between molecules.

d) Solids are usually more dense than liquids because the particles are touching all of their neighbours in a solid, but not in a liquid. This means that more mass/particles are concentrated into a given volume.

e) Particles are widely spaced.
Particles are moving very quickly. ↄ 90

2. a) Brownian motion is caused by the particles of ash/smoke being hit…
by molecules of air…
that are too small to be seen…
but moving very quickly.

b) This process is called diffusion.
The particles of the air freshener are hit by air particles…
which causes them to move and spread out.

c) The particles in a liquid are closer together.
The particles in a liquid are more closely packed so it is difficult for chemical particles to pass through. ↄ 91

3. a) (i) D
(ii) Solid

b) E or C

c) B or F

d) E

e) A

f) A or D ↄ 96

4. a) Any three from the following…
• You cannot reverse the process.
• There is a colour change.
• New substances are produced.
• There is a large energy change.

b) It is easy to reverse.
No new substances are made.

c) It stays the same.
No atoms are made or destroyed.
No substances are gained or lost from the egg. ↄ 97

Chapter 9

1. a)

(2 marks awarded for 2 or 3 correct lines on the left.
1 mark for only one correct line on the left. 2 marks

awarded for 2 or 3 correct lines on the right. 1 mark for only one correct line on the right.) ↄ 103

b)

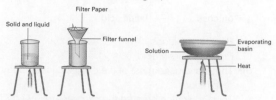

Any six marks from the following…
Correctly labelled diagram for each stage.
Add water…
to dissolve the salt.
Filter…
to remove the sand.
Heat…
to remove the water.

c) Distillation.

d)

The water is heated
…and then condensed. ↄ 104

2. a) So that people don't get poisoned/harmed. ↄ 102

b) C E A D F B
(All correct scores 5. One in wrong place scores 4.
Two in wrong place scores 3. Three in wrong place scores 2.) ↄ 106

c) Heat is up.
It should boil at exactly 100°C / It should leave no solid residue. ↄ 107

Chapter 10

1. a) Calcium carbonate →
… calcium oxide + carbon dioxide ↄ 110

b) $CaCO_3$ →
…$CaO + CO_2$ ↄ 111

c) Thermal decomposition. ↄ 115

d) Exothermic. ↄ 113

e) Calcium carbonate + hydrochloric acid →
…calcium chloride + water + carbon dioxide ↄ 111

f) Limewater.
Goes cloudy.
(Not puts out a burning splint because this is not a positive test for carbon dioxide.) ↄ 123

g) Something that speeds up a reaction
…without being used up/changed. ↄ 117

h) Independent variable (would change): the catalyst added.
Dependent variable (would measure): the time taken / speed of reaction.
Same volume/amount of acid.
Same mass/amount of limestone/calcium carbonate.
Same temperature. ↄ 7

2. a) When something reacts with oxygen or loses hydrogen / loses electrons. ↄ 116

b) Amount of water in can.
Amount of fuel burned.
Time of burning.
Starting temperature of water.
Height of can above flame.
Same apparatus/can. ↄ 7

c) 19°C

d)

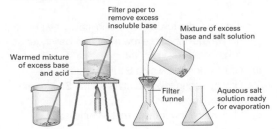

Marks awarded for the following…
- Bar chart chosen.
- Fuel on the *x* axis and temperature **change** on the *y* axis.
- Axes labelled.
- Units for temperature.
- Height of bars plotted correctly.
- Ruler used.

e) Butanol…

…because it had the largest temperature change.

f) To reduce heat loss.

Chapter 11

1. a) Sulfuric acid. ↩ 124

b) Sulfuric acid: any value below 6

Copper sulfate: 7 ↩ 121

c) Any four from the following…
- Measure sulfuric acid using a measuring cylinder.
- Put acid into beaker.
- Warm the acid over a gauze/tripod using a Bunsen.
- Add the copper oxide powder.
- Stir with glass rod/spatula.
- Filter unreacted copper oxide.
- Test solution using universal indicator. ↩ 125

d) Heat solution

Evaporate water…

…until half the solution has evaporated.

Leave to form crystals. ↩ 125

Filter paper to remove excess insoluble base

Mixture of excess base and salt solution

Warmed mixture of excess base and acid

Filter funnel

Aqueous salt solution ready for evaporation

e) C ↩ 9

2. a) Acids are 0–6.

Neutral is 7.

Alkalis are 8–14. ↩ 121

b) pH 1 = red.

pH 7 = green.

pH 14 = purple/blue. ↩ 12

Chapter 12

1. a) Units / °C

b) Larger temperature rise than for the other metals / calcium.

Very rapid fizzing. ↩ 133

c) Barium + hydrochloric acid →

…barium chloride + hydrogen ↩ 133

d) Burning splint…

…goes pop / explodes ↩ 123

2. a) Döbereiner grouped elements of similar properties

…and called them triads.

Newlands spotted a repeating pattern (with increasing atomic mass)

…and called this the law of octaves.

Mendeleev left gaps in his periodic table (for then undiscovered elements)

…and made predictions about their properties. ↩ 131

b) Any two from the following…
- At conferences.
- In journals/magazines.
- In books.
- Via the internet.
- Via email / letter / phone. ↩ 7

Chapter 13

1. a) Metal B.

Metal A.

Metal C.

Because metal B displaced both A and C, whilst metal C did not displace any other metal.

b) → A sulfate + C

c) Faster/more bubbles

Large temperature rise ↩ 138

2. a) Tin oxide + carbon →

…tin + carbon dioxide (or carbon monoxide).

b) Tin is less reactive than carbon.

c) Electrolysis.

d) Lots of…

…electrical energy is needed. ↩ 139

Chapter 14

1. a)

■ Oxygen

■ Silicon

■ Aluminium

■ Iron

■ Other elements

Correctly sized segments.

Key to colour/labelled segments.

Ruler used. ↩ 11, 143

b) Coal / Oil / Gas ↩ 143

c) Hard to extract from sand / silicon oxide. ↩ 143

d) Earthquakes/Tsunamis ↩ 148

Volcanoes.

2. a) Rock with large crystals cooled slowly…

…underground / intrusive.

Rock with small crystals cooled quickly…

…on the surface / under water.

b) Sedimentary.

c) Metamorphic. ↩ 145

3. a) 0.03 to 0.04%. ↩ 147

b) Photosynthesis.

Dissolving in the oceans. ↩ 147

c) Respiration/Volcanoes ↩ 147

d) (i) More humans means faster consumption of fossil fuels / increased energy demand etc.

(ii) Cutting down trees reduces photosynthesis.

(iii) Farming leads to the release of methane from cows. ↩ 148

e) Walk/cycle more instead of using the car/bus.

Use less electricity.

Turn central heating down. ↩ 148

f) More evidence/data.

Better data. ↩ 148

g) (Either of the following answers is acceptable but the mark is for the justification, not the yes/no.)

Yes, because the temperature is increasing and there is more freak weather. ↩ 148

Answers to practice questions

No, because there is not enough evidence to prove that the increase in temperature is actually causing the freak weather.

Chapter 15

1. a) Energy is transferred from the chemical store of the gas (and oxygen)

...and into the thermal store of the butter (and surroundings). ↺ 153

b) Energy = 2 × 1.5 = 3kW

Cost = 3 × 15 = 45p ↺ 157

c) Reflects...

infrared/radiation. ↺ 160

d) Power = $\frac{energy}{time}$

= $\frac{45,000}{30}$ = 1,500 watts ↺ 157

e) Heat is lost by convection currents.

A lid prevents convection currents / hot air/steam from escaping. ↺ 160

2. a) Vibrating particles

...bump into their neighbours / pass on vibrations to neighbours. ↺ 159

b) The air/gas is a poor conductor of heat. ↺ 160

c) Convection.

Still air stops convection currents. ↺ 160

d) Radiation.

Shiny surfaces are better at reflecting infrared/radiation.

3. a) Will eventually run out / are being made very slowly over millions of years. ↺ 155

b)

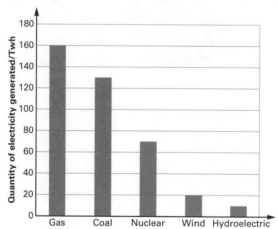

Bar chart chosen.

Type of energy resource on x axis and quantity of electricity on y axis.

Units on y axis.

Bars at correct height and drawn with a ruler. ↺ 11

c) Coal and gas are non-renewable.

Burning coal and gas releases carbon dioxide / causes global warming. ↺ 155

d) Nuclear fuels are non-renewable.

Very expensive to build.

Concerns over safety/radioactivity/health. ↺ 155

e) Advantages...

...renewable.

...UK is a windy place.

...no CO_2 emissions.

Disadvantages...

...not always windy.

...can spoil views. ↺ 155

4. a) Carbohydrates (sugar/starch).

Fats.

b) Carbohydrates: pasta, bread, potatoes, rice, sweets, fruit, sugar, etc.

Fats: butter, cheese, red meat, oil, etc.

c) Change the type of food burned.

Measure the temperature of the water before and after burning the fuel.

Keep the distance from the food to the boiling tube the same.

Keep the mass of the food the same.

Keep the volume of water the same.

Keep the starting temperature of the water the same.

Process results by calculating the temperature rise of the water.

The food that causes the largest temperature rise is the one that contains the most energy (if equal masses of food were burned). ↺ 156

Chapter 16

1. a)

(Length of arrows must be equal.) ↺ 168

b)

(Downward arrow must be identical to the answer to part (a) but upward arrow must now be longer or thicker.) ↺ 168

c) Gravity acting towards Earth decreases.

Gravity acting towards Mars (or the Sun, or other planets) increases. ↺ 168

d) Mars has a lower mass (not weighs less). ↺ 172

2. a) A and D ↺ 168

b) 6 km is 6,000m

1 min and 30s is 90 s

Speed = $\frac{6000}{90}$ = 67 m/s ↺ 164

c) Distance = speed × time

= 100 × 20 = 2000 m (or 2 km) ↺ 164

d) Area of four tyres = 4 × 0.1 m² = 0.4 m²

Pressure = $\frac{force}{area}$

Pressure = $\frac{6500}{0.4}$

= 16,250 N/m² ↺ 173

e) Force = pressure × area

Force = 250000 × 1.5

= 375,000 N ↺ 173

Chapter 17

1. a) 20 to 20 000 hertz

b) Wavelength: metres.

Frequency: hertz.

c) Loudspeaker: makes sound waves by converting electrical signals into vibrations of its diaphragm.
Microphone: detects sound waves which vibrate a diaphragm inside it and converts to an electrical signal.

d) Cleaning jewellery or watches.
Taking images of unborn babies.
Treating some injuries.
SONAR.

e) Sounds travel fastest in solids.
The particles are closest together.
So waves of vibrations are passed on most easily.

f) Sounds are waves of vibrating particles.
There are no particles in the vacuum of space.
So sounds cannot travel. ↺ 179

2. a) Canal / Drum / Ossicle bones / Cochlea / Auditory nerve

b) Transfers the vibrations from the sound on to the ossicle bones.

c) Loud sounds can puncture the ear drum, making hearing difficult.
Loud sounds can also flatten the hairs in the cochlea, making hearing difficult. ↺ 180

3. a) The distance that light travels in one year.

b)

c) Specular reflection is regular and comes from flat objects.
Diffuse reflection is irregular and comes from rough objects.

d) You see lightning before you hear thunder.

e) You see the red lettering because all other colours are absorbed and only red light is reflected into your eyes.
You see the white sign because all colours are reflected into your eyes.
A red filter would only allow red light through.
The box would look red because only red light would reach your eyes.
The writing would look black because no light would reach your eyes. ↺ 183

4. a) Pupil / Lens / Ciliary muscles / Retina / Optic nerve

b) The coloured muscle in the front of your eye which contracts and relaxes to control the size of your pupil.

c) They wear glasses because the lenses in their eyes cannot refract light enough to focus on their retina.
The lens in the glasses helps with this. ↺ 185

Chapter 18

1. a) Open switch / Motor / Voltmeter / Resistor

b)

Switch A	Switch B	Bulb
Up	Down	Off
Up	Up	On
Down	Up	Off
Down	Down	On

c) A measure of how much electrical energy is carried by the current in a circuit.

d) First voltmeter: 0 volts.
Second and third voltmeters: 12 volts. ↺ 191

2. a) Two insulators are rubbed together.
Electrons transfer from one to the other.
This causes a build-up of static electricity.

b) If the rods both had the same charge they would repel.
If the rods had opposite charges they would attract. ↺ 194

3. a) Domains.

b) An unmagnetised piece of iron, nickel or cobalt is turned into a temporary magnet when inside another magnetic field.
Permanent magnets are made by stroking a piece of iron, nickel or cobalt with another permanent magnet to align the domains.

c)

d) The poles.
This is where magnetic field lines are closest.

e)

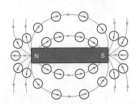

f) Steel is an alloy of iron. ↺ 196

4. a) They can easily be turned on and off.

b)

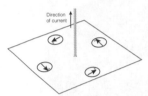

c) A magnet is moved through a coil of wire.
Or a coil of wire is rotated in a magnetic field. ↺ 197

Chapter 19

1. a) Mercury and Venus.

b) Sun: luminous.
Mercury: non-luminous.
Venus: non-luminous.
Earth: non-luminous.
Moon: non-luminous.

c) The objects are too close to each other for their size.
(The scale of the distances between the planets is not correct.)

d) The right hand side should be shaded.

e) The four planets nearest the Sun are all made from rock.
They are also the warmest and smallest of the planets.
The four planets furthest from the Sun are made from gases.
They are also the coldest and largest of the planets. ↺ 203

2. a) weight (N) = mass (kg) × gravitational field strength (N/kg)

b) The Sun has a larger mass than the Earth so gravitational field strength would be larger.
This would make your weight increase.
The Moon has a smaller mass than the Earth so gravitational field strength would be less.
This would make your weight decrease.

c) There is no air in space (the Moon has no atmosphere).
There is no air resistance to slow the feather more than the hammer.
So both will accelerate due to gravity at the same rate. ↺ 205

1. **This question is about plant and animal cells.**

   (a)

   A plant cell from a leaf

   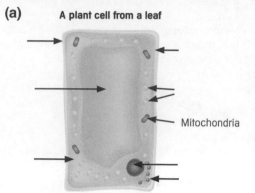

   Animal cell

   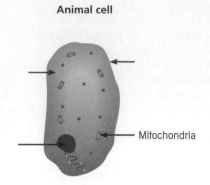

   Mitochondria

   These diagrams show an animal and a plant cell. Label the components that are present in both cells. The first one has been completed for you. *(3 marks)*

   (b) State the function of mitochondria. *(1 mark)*

   (c)

   State the name of this cell and describe its shape. *(2 marks)*

   (d) Describe the adaptation of this cell to its function. *(2 marks)*

   (e) Explain why both animal and plant cells have mitochondria but only plant cells have chloroplasts. *(4 marks)*

2. **This question is about ecosystems.**

   (a) State what bioindicators are. Give an example of one in your answer. *(2 marks)*

   (b) Describe why food webs are an example of independency. *(1 mark)*

   (c)

   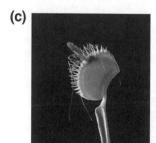

   The Venus flytrap plant catches insects, which land on its trap-like leaves. Explain why this makes them both a producer and a consumer. *(2 marks)*

   (d) Suggest why plants like this have evolved to catch insects. *(1 mark)*

3. **Seb is making fertilisers. Most fertilisers are salts that contain nitrogen. They often contain potassium and phosphorus as well.**

    (a) Use the periodic table to find the chemical symbol for potassium. *(1 mark)*

    (b) Ammonium nitrate is a very important fertiliser.
    A bag of ammonium nitrate has this hazard symbol on it.

    State the meaning of this hazard symbol and describe a suitable safety precaution that Seb should take when using ammonium nitrate that is specific to this hazard symbol. *(2 marks)*

    (c) Ammonium nitrate is made by reacting ammonium hydroxide (an alkali) with a specific acid. Name the acid needed to make ammonium nitrate. *(1 mark)*

    (d) Describe how Seb should produce a pure dry sample of ammonium nitrate. Include in the method the safety precautions for the chemicals and the procedures and details of how Seb should make sure that the sample is neutral. *(5 marks)*

    (e) A fertiliser can be made in a neutralisation reaction between potassium hydroxide and sulfuric acid. Write a word equation for this reaction. *(2 marks)*

4. **This question is about the structure of the atom.**

    (a) What name is given to the central part of an atom, which contains protons and neutrons?

    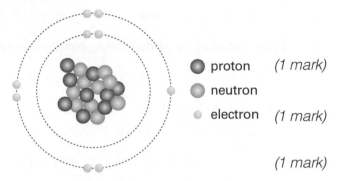

    (b) What is the atomic number of this element?

    (c) Use the periodic table to find the name of this element. *(1 mark)*

    (d) Fill in the table below.

Particle	Mass	Charge
Proton		
Neutron		
Electron		

(6 marks)

5. **This question is about electricity.**

(a)

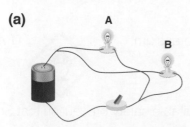

The circuit above has two identical bulbs. Draw on two switches and label them 1 and 2. Switch 1 will control the both bulbs and switch 2 will control bulb B only.   *(2 marks)*

(b) State what type of circuit it is.   *(1 mark)*

(c) A voltmeter is placed across bulb A and reads 6 volts. State what the reading across bulb B will be.   *(1 mark)*

(d) An ammeter is placed before the circuit branches and reads 3 amps. Draw the correct symbol for an ammeter at another point in the circuit where it will read the same.   *(2 marks)*

(e) The ammeter is moved to just before bulb A and reads 1.5 amps. State the reading it would show if it were moved just after bulb B.   *(1 mark)*

6. **This question is about stretching springs.**

(a) State how energy is stored in a spring.   *(1 mark)*

(b) Gemma decides to investigate how the length of a spring is affected by the force used to stretch it. What is the independent variable in her investigation?   *(1 mark)*

(c) Gemma investigates two springs and plots a graph of her results. Write a conclusion from her results.   *(2 marks)*

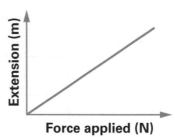

(d) Suggest why Gemma should not apply very large forces to stretch the spring.   *(1 mark)*

(e) Gemma repeats the experiment with a different spring. This time, the line is steeper when plotted on the same graph. Which spring is stiffer, the first or second one she tested? Explain your answer.   *(2 marks)*

7. **This question is about rocks and the rock cycle.**

(a) Igneous rocks are formed from volcanoes. Sometimes these rocks are formed underground from magma and sometimes they are formed on the surface from lava. How could you deduce where an igneous rock had formed from its structure?   *(3 marks)*

(b) Describe how sedimentary rocks are formed from igneous rocks.   *(2 marks)*

(c) Slate is a metamorphic rock formed from shale. Describe how metamorphic rocks are formed.   *(2 marks)*

(d) Only one type of rock can contain fossils. State which type of rock it is and explain why fossils cannot be found in the other two types. *(3 marks)*

(e) The formation of igneous and metamorphic rocks is caused by movements of the tectonic plates that make up the crust of the Earth. What causes these plates to move? *(2 marks)*

8. **This question is about keeping drinks hot and cold.**

(a) A steel flask is often used to keep drinks warm in cold weather. It has a double lining with a vacuum between the two shiny metal surfaces.

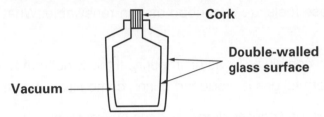

Explain how this flask prevents the hot drink from cooling by conduction and convection. *(2 marks)*

(b) Explain how this flask prevents the hot drink from cooling by radiation. *(1 mark)*

(c) The flask can also be used to keep a drink cold. Explain why. *(1 mark)*

(d) Drinks can also be kept cool by putting ice cubes in them. Describe the heat transfer process when an ice cube melts in a drink. *(1 mark)*

(e) When a drink is very hot, we often blow on it to cool it down. Explain how this works. *(2 marks)*

9. **This question is about sound waves.**

(a) State the type of waves that includes sound waves. *(1 mark)*

(b) State the name of sounds above 20 000 hertz. *(1 mark)*

(c) Explain why sound waves travel faster underwater than in the air. *(2 marks)*

(d) Fishermen can use SONAR to identify shoals of fish. Describe how they do this. You can draw longitudinal sound waves to help you. *(3 marks)*

(e) State the equation that could be used to calculate how far away the shoal of fish is. *(1 mark)*

10. **This question is about the solar system and beyond.**

(a) State what a galaxy is. *(1 mark)*

(b) Our solar system is made up of the Sun, planets and many moons that surround them. Explain how gravity affects these objects. *(2 marks)*

(c) The gravitational field strength of the Moon is 1.6 N/kg. The astronaut has a mass of 75 kg. Calculate their weight on the Moon. *(2 marks)*

(d) Explain how and why this number would change when the astronaut returned to Earth. *(2 marks)*

(e) Explain why the Earth has seasons. *(3 marks)*

11. **This question is about burning fuels.**

   (a) Burning fuels can be described as an oxidation reaction. What is meant by the term oxidation?

   (1 mark)

   (b) How well a fuel burns depends on the supply of oxygen. This can be seen in a Bunsen burner with the air hole open or closed. State the type of combustion that happens with a good supply of oxygen and describe two advantages of this type of combustion.

   (3 marks)

   (c) Much of the energy used by people in developed countries is released by burning fossil fuels. These fuels are described as non-renewable. What is meant by this term?

   (1 mark)

   (d) Fossil fuels often contain sulfur as an impurity. If this is not removed from the fuel before it is burned, a harmful gas is produced during combustion. Name this gas. *(1 mark)*

   (e) Describe how this gas causes damage to the environment. *(1 mark)*

12. **This question is about food and your digestive system.**

   (a) A balanced diet contains the correct amount of the six food groups.
   State the names of two of these groups. *(2 marks)*

   (b) Describe what can happen if your diet is not healthy. *(2 marks)*

   (c) There are millions of villi in your digestive system. Describe how they are adapted to their function. *(4 marks)*

   (d) Explain why we have a digestive system. *(1 mark)*

   (e) You are prescribed some antibiotics by your doctor to kill bacteria that are infecting your large intestine, which is not working properly. Describe what symptoms you might expect and explain why you might have them. *(2 marks)*

13. **This question is about reproduction in plants and animals.**

   (a) State the function of the scrotum. *(2 marks)*

   (b) Describe an adaptation of a human sperm and ovum. *(2 marks)*

   (c) Explain why mammals have a regular menstrual cycle. *(2 marks)*

   (d)

   A    B    C

   Diagrams A and B above are of seeds. Diagram C is a fruit. Describe the methods of seed dispersal of A and B, which are the same, and then C. *(2 marks)*

(e) The flower is the reproductive organ of many plants. The male part of the flower is called the stamen. It consists of the anthers and filaments. The female part of the flower is called the carpel. State the three names of the parts that make up the carpel. *(3 marks)*

14. **This question is about a compound called cobalt chloride.**

(a) These symbols as found on a bottle containing cobalt chloride crystals. What do they mean?

A   B   *(2 marks)*

(b) Cobalt chloride is a red solid that dissolves in water to produce a pink solution. State the solvent and the solute in a solution of cobalt chloride. *(2 marks)*

(c) Describe how to produce a sample of pure and dry cobalt chloride crystals from a solution of cobalt chloride. Include the apparatus needed in your answer. *(3 marks)*

(d) Explain how the solution could be separated in a way that will allow the water to be collected. *(3 marks)*

(e) How could the water be tested to check that it is pure? Describe the expected result. *(3 marks)*

15. **This question is about skeletons and muscles.**

(a) There are five common types of joints. State what type of joint they are. *(2 marks)*
(b) Describe two functions of your skeleton. *(2 marks)*

(c) Muscles are tissues that are found in most animals. State the definition of a tissue. *(1 mark)*

(d) Describe how muscles can help you move. *(4 marks)*

(e) Explain how antagonistic muscles work and why they come in pairs. You may draw a diagram to help you. *(4 marks)*

16. **This question is about burning magnesium.**

    (a) When magnesium burns in air, it reacts with oxygen. Write a word equation for this reaction. *(2 marks)*

    (b) The reaction is very exothermic. Explain what this means. *(1 mark)*

    (c) Look at the table of data below. Describe the trend in the results.

Mass of magnesium burned (g)	Mass of magnesium oxide produced (g)
1.2	2.0
2.4	4.0
3.6	6.0
4.8	8.0

    (1 mark)

    (d) Which type of graph would be best for representing these data?

    Bar chart   pie chart   line graph *(1 mark)*

    (e) Predict the mass of magnesium oxide that would be made from 0.6 g of magnesium. *(1 mark)*

17. **Xander is playing in a playground. This question is about the forces that are acting on him and their effects.**

    (a) Describe the forces that are acting on Xander when he is sitting in a stationary swing. *(2 marks)*

    (b) Xander gets a friend to give him one big push and he starts swinging. After a while, he slows down and stops. Explain why. *(1 mark)*

    (c) The roundabout is moving at a constant speed, with Xander sitting near the edge. Is there a force on Xander? Explain your answer. *(2 marks)*

    (d) On the see-saw, Xander sits 2 m away from the pivot. He weighs 600 N. Calculate the size of his moment. Give the unit with your answer.

    Moment = force × distance from pivot *(2 marks)*

    (e) Archie is lighter than Xander. He must sit 3 m from the pivot. How much does Archie weigh? *(2 marks)*

18. **This question is about energy stores and transfers.**

    (a) State the main way that energy is stored in these objects.

    (i) A moving car.

    (ii) A mixture of a fuel and oxygen.

    (iii) A rollercoaster at the highest point on its track.

    (iv) A hot iron. *(4 marks)*

(b) In a farm on a remote island, a wind turbine is used to generate electricity to power a heater, which warms the air in the house. Draw an energy transfer diagram to represent this process. *(2 marks)*

(c) The battery in a toy car transfers 360 joules of energy to the motor in 1 minute. The motor transfers 60 joules to the kinetic store of the car. The rest is transferred to the thermal store of the car and surroundings. Draw a labelled Sankey diagram to represent this process. *(3 marks)*

(d) Calculate the power of the motor, using the total amount of energy it transfers. Include the unit in your answer.

Power = energy / time *(2 marks)*

(e) How much energy is transferred if the motor runs for 5 minutes? *(3 marks)*

19. **This question is about hydrogen fuel cells. Hydrogen fuel cells allow electricity to be produced from the reaction between hydrogen (H_2) and oxygen (O_2). The only product of this reaction is water.**

(a) The word equation for the reaction in a hydrogen fuel cell is...

Hydrogen + oxygen → water

Write down the name of a reactant in this equation. *(1 mark)*

(b) The symbol equation for the reaction is...

$H_2(g) + O_2(g) → H_2O(l)$

Balance the symbol equation. *(1 mark)*

(c) State the meaning of the terms (g) and (l) in the equation above. *(2 marks)*

(d) Suggest why the hydrogen fuel cell is described as being more environmentally friendly than a car that runs on fossil fuels. Explain your answer. *(2 marks)*

(e) Hydrogen for use in fuel cells can be obtained from water using electrolysis. This means that electricity is used to split water up into hydrogen and oxygen. Suggest why this might still involve the release of carbon dioxide into the atmosphere. *(1 mark)*

20. **This question is about breathing and your gas exchange system.**

(a) State where your body has rings of cartilage. *(1 mark)*

(b) Describe the effects that asthma has upon your gas exchange system. *(2 marks)*

(c) There are millions of villi in your lungs. Describe how they are adapted to their function. *(4 marks)*

(d) Describe why oxygen diffuses into your blood from your alveoli. *(3 marks)*

(e) Explain why people who smoke are often more tired at the end of a run. *(6 marks)*

21. **This question is about heredity and evolution.**

    (a) State the term used to describe the offspring of asexual reproduction. *(1 mark)*

    (b) Describe the structure of DNA. *(3 marks)*

    (c)

Blood group	Percentage of UK population
A	42
B	10
AB	4
O	44

    The table shows the percentage of the UK population with the four types of blood group. Draw a graph to show this information. *(5 marks)*

    (d) Explain why genetic diversity is important. *(2 marks)*

    (e) MRSA bacteria have evolved to not be killed by antibiotics used to treat many other infections. They are called antibiotic-resistant. Explain how they have evolved. *(5 marks)*

22. **This question is about photosynthesis and respiration.**

    (a) State the equation for respiration. *(2 marks)*

    (b) Describe the differences between respiration and photosynthesis. *(4 marks)*

    (c) Describe the similarities and differences between respiration and fermentation. *(4 marks)*

    (d) Explain why your breathing rate increases during exercise. *(3 marks)*

    (e) A biosphere is a sealed container in which plants and animals can survive. Explain what would happen and why if the following were placed inside:
    – Plants alone.

    – Animals and plants together.

    – Animals alone. *(6 marks)*

23. **This question is about the properties and uses of elements.**

    The table of data will help you to answer parts of this question.

Metal	Relative density	Relative strength	Relative electrical conductivity	Reactivity	Cost (£/kg)
Steel	8	8	10	Medium	1
Platinum	21	5	9	Very low	60,000
Copper	9	5	50	Low	3
Aluminium	3	6	30	Low	0.5
Titanium	4	10	25	Low	17

(a) Suggest why the majority of titanium extracted in the world is used to build aeroplanes. *(2 marks)*

(b) Which metal is used for electrical wiring? Explain why. *(2 marks)*

(c) Platinum is used for jewellery. Explain why. *(1 mark)*

(d) The metal used for high-voltage electrical power cables is different from the metal used for wiring in homes. Suggest which metal is used for high-voltage cables and explain your answer. *(3 marks)*

(e) You are given a sample of a pure element and must decide if it is a metal or a non-metal. Describe what tests you would do and what you would expect to find if it was a metal. *(4 marks)*

24. **This question is about light waves.**

(a) State the law of reflection. *(1 mark)*

(b)

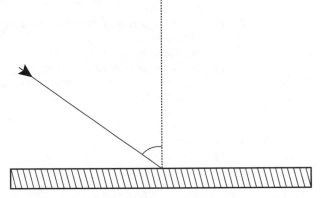

A ray of light moves towards a flat mirror. Continue the arrow to show the direction of this ray. Draw on the normal line and label the angles of incidence and reflection. *(6 marks)*

(c) Describe diffuse reflection and give an example in your answer. *(2 marks)*

(d)

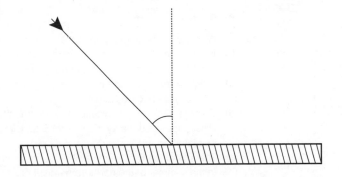

A ray of light moves from air into a block of glass and out the other side. This is called refraction. Continue the arrow to show the direction of this ray. *(5 marks)*

(e) Describe the process of dispersion and explain why it occurs. *(4 marks)*

1. a) Nuclei.
   Membrane.
   Cytoplasm. ⟲ 18
   b) The site of respiration. ⟲ 19
   c) Red blood cell.
   Biconcave. ⟲ 20
   d) It has a biconcave shape and no nucleus.
   This maximises the surface area.
   To absorb the maximum amount of oxygen. ⟲ 20
   e) Both plant and animals respire.
   So both need mitochondria.
   Animals do not photosynthesise.
   So animal cells do not need chloroplasts. ⟲ 20

2. a) Organisms whose presence tells us about the ecosystem.
   Bloodworms, sludgeworms, stonefly nymphs or lichens. ⟲ 73
   b) All the organisms depend upon each other. (Or if one organism suffers then this can affect the others.) ⟲ 72
   c) Producer: Venus fly traps have green leaves that photosynthesis.
   Consumer: Venus fly traps eat insects. ⟲ 73
   d) The soil they live in does not have enough nutrients for them.
   They are at an advantage catching insects over those plants that don't. ⟲ 71

3. a) K. ⟲ 132
   b) Oxidising agent.
   Keep away from fuels/anything that might burn. ⟲ 9
   c) Nitric acid. ⟲ 123
   d) Ammonia solution/ammonium hydroxide – is an irritant/corrosive.
   Nitric acid is an irritant – so avoid contact with skin.
   Add nitric acid to the ammonium hydroxide.
   Check it is neutral using universal indicator.
   Heat the solution to evaporate the water. ⟲ 125
   e) Potassium hydroxide + sulfuric acid →
   ...potassium sulfate + water.

4. a) Nucleus. ⟲ 95
   b) 9. ⟲ 96
   c) Fluorine. ⟲ 132
   d) ⟲ 96

Particle	Mass	Charge
Proton	1	+ (positive)
Neutron	1	neutral/0
Electron	0/zero/almost zero	– (negative)

5. a) Switch one is before or after branch.
   Switch two is on branch with bulb B. ⟲ 192
   b) Parallel. ⟲ 192
   c) 6 volts. ⟲ 193
   d)

   —(A)—

   Position must be after branches have joined again but before the return to the cell. ⟲ 193
   e) 1.5 amps. ⟲ 192

6. a) Elastic store. ⟲ 152
   b) Force. ⟲ 170
   c) As the force increases, the extension increases. ⟲ 170
   d) It could damage the spring/could be dangerous. ⟲ 170
   e) Second spring.
   It takes more force to produce the same extension/has lower extension for the same force. ⟲ 170

7. a) Look at the size of crystals.
   Large crystals – cooled slowly underground.
   Small crystals – cooled quickly on the surface. ⟲ 145
   b) Igneous rock is weathered/eroded and small pieces of rock are transported by rivers to the sea.
   Settle on bottom, pressure. ⟲ 145
   c) High temperatures.
   Pressure. ⟲ 145
   d) Sedimentary rocks can contain fossils.
   Igneous rock cannot because no living thing can survive in molten rock and any existing fossils in sedimentary rock will be destroyed when it is melted to form igneous rock.
   Metamorphic rock cannot because fossils will be destroyed by heat and pressure. ⟲ 145
   e) Convection currents.
   In the mantle. ⟲ 142

8. a) Conduction and convection both need particles/substance/matter.
   No particles in vacuum. ⟲ 160
   b) Shiny surfaces reflect infrared/radiation. ⟲ 160
   c) Stops thermal energy from entering the flask. ⟲ 160
   d) Thermal energy moves from the drink into the ice cube. ⟲ 159
   e) Moving air – increases convection. ⟲ 159

9. a) Longitudinal. ⟲ 177
   b) Ultrasound. ⟲ 180
   c) (Sounds travel faster in liquids than gases.)
   The particles are closer together.
   Vibrations are passed on more easily. ⟲ 182
   d) Waves spread out from the boat.
   Rays that reach the fish are reflected back to the boat.
   These rays reach the boat faster.
   Rays that don't reach the fish carry on past them.
   (They can be reflected from the seabed.) ⟲ 182
   e) distance = $\dfrac{\text{speed}}{\text{time}}$ ⟲ 182

10. a) A galaxy is a large group (millions) of solar systems. ⟲ 203
    b) It holds the planets in orbit around the Sun.
    It holds the moons in orbit around the planets. ⟲ 205
    c) weight = mass × gravitational field strength
    120 N = 75 kg × 1.6 N/kg ⟲ 205
    d) There is a stronger gravitational field strength on Earth.
    (Their mass would stay the same.) But their weight would increase. ⟲ 205
    e) The axis of the Earth is on a tilt.
    In the summer of the northern hemisphere it is tilted towards the Sun.
    It is closer to the Sun.
    So it is hotter. (Or the reverse.) ⟲ 204

11. a) When something gains oxygen in a reaction. ⟲ 116
    b) Complete combustion.
    Releases more energy.
    Doesn't produce poisonous carbon monoxide/dirty soot. ⟲ 114
    c) It will run out one day. ⟲ 155
    d) Sulfur dioxide. ⟲ 75, 155
    e) Acid rain. ⟲ 75, 155

12. a) Vitamins/Dietary fibre/Minerals/Proteins/Carbohydrates/Lipids (2 marks for 2 correct examples) ⟲ 38

b) Weight gain or obesity or related medical condition such as heart disease.
Weight loss, malnutrition or starvation. ⟲ 40

c) They increase the surface area of the small intestine.
They have a large blood supply.
They have walls one cell thick.
This increases the amount of digested food absorbed into the blood. ⟲ 31

d) The large insoluble food we eat needs to be broken down.
Before it can be absorbed into the blood. ⟲ 30

e) Diarrhoea.
Your large intestine absorbs water and if it isn't working properly then your faeces is likely to have too much water in it. ⟲ 32

13. a) To hold the testes away from the body.
To keep the temperature lower to maximise sperm production. ⟲ 49

b) Sperm: small and streamlined to swim quickly to the ovum.
Sperm: contains enzymes in its head to enter the ovum. ⟲ 51
Ovum: contains an energy store for the new organism.

c) To renew the lining of the uterus each month.
In case a fertilised ovum implants to grow into a baby. ⟲ 50

d) A and B: wind.
C: eaten by animals. ⟲ 57

e) Stigma.
Style.
Ovary. ⟲ 56

14. a) A = Dangerous for the environment.
B = Toxic. ⟲ 9

b) Solvent = water.
Solute = cobalt chloride. ⟲ 103

c) Place the solution in an evaporating basin. Also need a tripod and gauze.
Heat using a Bunsen burner.
Allow the water to evaporate. ⟲ 104

d) Use distillation.
Use a condenser – to cool the water vapour down. ⟲ 105

e) Either…
Heat until it boils – and check the temperature – which should be exactly 100°C.
Or…
Heat until it boils – and keep heating until it has all evaporated – and there should be no residue. ⟲ 107

15. a) Slightly moveable: spine.
Ball and socket: hip and shoulder.
Hinge: knee and elbow.
Pivot: neck.
Fixed: skull. ⟲ 28

b) Supports muscles and internal organs.
Protects internal organs.
Moves.
Produces red and white blood cells. ⟲ 28

c) Lots of identical, specialised cells in the same place completing the same function. ⟲ 21

d) Muscles are connected to bones.
By tendons.
Muscles can contract and relax.
When they contract they pull on tendons which pull on bones.
Bones act as levers. ⟲ 29

e) Muscles can only contract and relax.
So they need to work in pairs.
When one contracts the other relaxes.
When the second contracts the first relaxes.
(Suitable marks awarded for drawings.) ⟲ 29

16. a) Magnesium + oxygen → … ⟲ 110
…magnesium oxide.

b) It releases energy. ⟲ 113

c) As the mass of magnesium burned increases, so does the mass of magnesium oxide. ⟲ 12

d) Line graph. ⟲ 11

e) 1 g. ⟲ 12

17. a) Gravity/weight acts downwards.
Tension in the chains of the swing acts upwards. ⟲ 168

b) Air resistance/friction. ⟲ 154

c) Yes.
He is changing direction. ⟲ 168

d) Moment = 600 N × 2 m = 1200 Nm. ⟲ 171

e) $\dfrac{1200}{3}$
= 400 N ⟲ 171

18. a) Kinetic store.
Chemical store.
Gravitational store.
Thermal store. ⟲ 152

b) Kinetic store →
…thermal store. ⟲ 153

c)

Correct labels.
Correct sized output arrows.
Correct energy labels with units. ⟲ 154

d) 6 watts (or 6 J/s). ⟲ 157

e) Energy = power × time
Energy = 6 × 300 seconds
1800 J ⟲ 157

19. a) Hydrogen/oxygen. ⟲ 110

b) $2H_2 + O_2 \rightarrow 2H_2O$ ⟲ 112

c) (g) means gas.
(l) means liquid. ⟲ 111

d) It produces only water/produces no carbon dioxide.
So it doesn't contribute as much to global warming. ⟲ 155

e) Fossil fuels may have been used to generate the electricity in a power station. ⟲ 155

20. a) Trachea. ⟲ 33

b) The lining of the airways become inflamed.
This makes it hard to breathe. ⟲ 43

c) They increase the surface area of the lungs.
They have a large blood supply.
They have thin walls.
This increases the amount of oxygen that diffuses into your blood.
Or this increases the amount of carbon dioxide that diffuses into your lungs. ⟲ 34

d) Diffusion is the movement of particles from an area of high to low concentration.
When you breathe in oxygen is at high concentration in your lungs.
And in a low concentration in your blood.
(So moves from your lungs to your blood) ⟲ 34

e) Smoking breaks down the walls of the alveoli.
This reduces the surface area of the lungs.
This makes it harder for oxygen to diffuse into the blood.
This means muscles cells have less oxygen.
This means muscles cells cannot respire as easily.
This means less energy is released. ⟲ 42

Answers to mixed test-style paper

21. a) Clone. ↻ 79

b) Double helix.
Made of four bases.
A-T and C-G (or T-A and G-C). ↻ 81

c) Bar chart drawn.
One mark for each correctly plotted point. ↻ 84

d) Without it a single event could kill all organisms.
Like disease or environmental change.
(The species cannot evolve without it.) ↻ 80

e) Variation exists in populations of all living organisms.
(So some bacteria are naturally resistant.)
Because the numbers of organisms of most populations
remain the same the individuals within them must
compete with each other to survive. (The bacteria
compete with each other.)
Variation results in some individuals possessing
characteristics that mean they are better adapted to
survive and reproduce than others. (The bacteria that
are antibiotic resistant have a huge advantage.)
Heredity means that these advantageous characteristics
are passed to their offspring. (The offspring formed from
antibiotic-resistant bacteria are also resistant.)
Over many, many generations small changes in these
characteristics can add up and new species of life are formed.
(This has already happened for these bacteria.) ↻ 85

22. a) ↻ 65

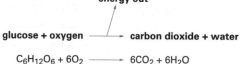

b) Photosynthesis occurs in plants.
Respiration occurs in all living cells.
Photosynthesis requires energy.
Respiration releases energy from glucose. ↻ 65

c) They both release energy from glucose.
They both produce carbon dioxide.
Respiration produces water.
Fermentation produces alcohol.
Fermentation only happens in some bacteria and yeast.
Respiration happens in all other organisms. ↻ 67

d) Your muscles are using more energy to move.
So they need to respire more.
So your body requires more oxygen.
(So your breathing rate needs to increase.) ↻ 66

e) Plants alone: the plants would respire producing carbon
dioxide, the plants would photosynthesis producing
oxygen; so they would live.
Animals and plants together: the animals and plants
would respire producing carbon dioxide, the plants
would photosynthesis producing oxygen; so they
would live.
Animals alone: the animals would respire, use all the
oxygen; so they would die. ↻ 65

23. a) It is strong
… and lightweight. ↻ 129

b) Copper.
It is a good conductor (of electricity). ↻ 129

c) It is unreactive (accept expensive).

d) Aluminium.
It is a good conductor…
… but it is lighter/cheaper/stronger. ↻ 129

e) Does it conduct electricity? Expect yes.
Does it have a high melting point? Expect yes. ↻ 128, 130

24. a) The angle of incidence always equals the angle of
reflection. ↻ 184

b)

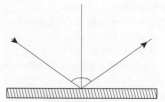

One mark for:
Straight lines.
One arrowhead on each line.
Normal line draw at right angles to mirror from point
where ray meets mirror.
Angle of reflected line the same as incident line (when
compared with the normal).
Angle of incidence is between the ray of incidence and
the normal line.
Angle of reflection is between the ray or reflection and
the normal line. ↻ 184

c) Light rays reflect from the surface of a rough object at
different angles.
Reflections from ripples on the surface of water. ↻ 184

d)

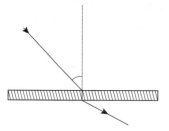

One mark for:
Straight lines.
One arrowhead on each line.
Change of direction as light enters the block (towards
the normal).
Change of direction as light exits the block (away from
the normal).
The direction of the line before and after the block is the
same. ↻ 185

e) White light is made from: red, orange, yellow, green,
blue, indigo and violet.
This refracts when it slows down. (This refracts when it
moves from air to glass or water.)
Red light refracts the least, orange a little more
(and so on).
So the colours separate. ↻ 187

Key word dictionary and index

Accurate   6
How close a measurement is to the true value.

Acid   121
Chemical that has a pH of less than 7.

Aerobic   65
In the presence of oxygen.

Alkali   121
A solution that has a pH of more than 7, made when a base dissolves in water.

Alkali metals   132
The elements in group 1 of the periodic table: lithium, sodium, potassium, rubidium and caesium.

Alloy   128
A mixture of two or more metals. (However, note that steel is an alloy made from iron with the non-metal carbon.)

Ammeter   191
An electrical component that measures current in amps, A.

Amnion   52
The sac that contains fluid that surrounds and protects an unborn baby.

Amniotic fluid   52
The liquid found within an amniotic sac that surrounds and protects an unborn baby.

Amplitude   178
The vertical distance that a wave travels from its starting position.

Anaerobic   65
In the absence of oxygen.

Angle of incidence   184
The angle at which a ray of light hits a mirror in a ray diagram.

Angle of reflection   184
The angle at which a ray of light reflects from a mirror in a ray diagram.

Antagonistic pair   29
Two muscles that work opposing each other. When one contracts the other relaxes.

Anther   56
The male part of a flower which holds the pollen.

Asexual reproduction   55
The formation of a new organism genetically identical to its one parent.

Asthma   43
A medical condition that causes swelling of airways in the gas exchange system.

Atom   90, 102
The smallest particle of an element. Atoms are made from protons, neutrons and electrons.

Atomic number   95
The number of protons in an atom of an element.

Attract   194
The coming together of two opposite electrical charges (positive and negative) or magnetic poles (north and south).

Auditory range   180
The normal range of human hearing; from 20 to 20 000 hertz.

Balanced diet   39
One with the correct amount of the six food groups necessary to live healthily.

Base   129
A substance that is chemically opposite to an acid, so it will neutralise an acid to make a salt. Many bases are metal oxides.

Base pair   81
Two molecules which exist in pairs (A-T and G-C) to make up DNA.

Battery   191
Two or more electrical cells joined together which can convert chemical energy into electrical energy in a circuit.

Bioaccumulation   73
The build-up of toxins in food chains that begin at low levels and become more concentrated at each trophic level.

Bioindicator   76
An organism that tells us about the level of air or water pollution present in an ecosystem.

Biomass   155
Material that makes up living or recently living organisms.

Biomechanics   29
The study of how bones and muscles work together.

Boil   93
When a liquid turns into a gas at the boiling point of the substance.

Bone   27
A rigid organ that forms the internal skeleton of vertebrates.

Conductor *128, 159*
A substance that is able to conduct either heat (thermal conductor) or electricity (electrical conductor).

Consumer *71*
A term given to animals that are above the producer (almost always plants) in food chains.

Contact forces *172*
A type of force that acts between two objects that are touching.

Continuous variation *84*
Differences found in organisms within a species that produce data which comes in a range.

Contract *29*
When muscles shorten for movement.

Contraction *53*
The rapid shortening of muscles in the uterus of a pregnant woman about to give birth.

Convection *159*
When thermal energy is transferred through a fluid (liquid or gas) as a result of the substance itself heating up, getting less dense and then rising away from the source of thermal energy.

Convex *186*
A lens that is thicker in the middle than at the edges.

Cornea *185*
Tissue found at the front of your eye which protects it and also refracts light on to your lens.

Crust *142*
The solid, outermost layer of the Earth, which is made up of tectonic plates.

Current *192*
The flow of charge around a circuit. Current is measured in amps, A.

Deciduous *63*
Plants which drop their leaves in autumn and regrow them in spring.

Deficiency disease *41*
An illness caused when your body does not have enough specific vitamins or minerals.

Density *128, 174*
A property of an object or substance which is calculated by dividing its mass by its volume.

Diffuse reflection *184*
When light waves bounce from a rough surface at a different angle from which they hit it.

Diffusion *31, 91*
The movement of particles of a substance from an area where there are lots of them (high concentration) to an area where there are fewer of them (lower concentration).

Digestion *30*
The breaking down of food and its absorption into your blood.

Digestive system *30*
The organs that break down food and absorb it into your blood.

Dilute *122*
When a solution contains a lot of solvent (typically water) and not much solute.

Discontinuous variation *84*
Differences found in organisms within a species that produce data which comes in groups.

Disperse (dispersal) *57*
The spreading of plant seeds by wind, water, animals or ejection.

Dispersion *187*
When light passes through a prism and is split into a spectrum of colours.

Displacement reaction *138*
When a more reactive element pushes a less reactive element out from a compound.

Dissolve *111*
When a solute (usually a solid) breaks into tiny particles (molecules or ions) and spreads out throughout the particles of a solvent (a liquid).

DNA *81*
(deoxyribonucleic acid) the basic unit of heredity or inheritance.

Domain *196*
Tiny regions which can be aligned in magnetic metals to make them into a magnet.

Double helix *81*
The shape of DNA, which looks like a twisted rope ladder.

Drug *44*
Any substance taken as medicine, to intoxicate or to enhance performance.

Drum *180*
The thin membrane in your ear that vibrates when sound waves pass down your canal which then vibrates your ear bones (ossicles).

Echo *180*
The reflection of a sound wave that is heard a short period of time after the original.

Ecosystem *71*
A community of living organisms (plants, animals and micro-organisms) and the environment in which they live.

Electric field *194*
The area in which an attracting or repelling force occurs.

233

Key word dictionary and index

Key word dictionary and index

Longitudinal 177
Waves like sound that are started by a movement in the direction of the wave.

Luminous 202
Objects that create light.

Machine 153, 158
An object that applies a smaller force through a bigger distance (or a bigger force through a smaller distance) for a certain amount of energy transferred.

Magnet 196
A piece of metal that produces a magnetic field.

Magnetic 196
A term used to describe a metal that has been turned into a magnet.

Magnetic field 152, 196
The area in which magnetic metals experience an attracting or repelling force.

Mantle 142
The semi-solid layer of the Earth underneath the crust, which moves very slowly owing to convection currents.

Mass 205
A measurement of the amount of matter (stuff) in an object. Mass is measured in kilograms, kg.

Melt 93
When a solid turns into a liquid when it is heated.

Menstrual cycle 50
The regular cycle, approximately every 28 days, that prepares a woman's uterus for pregnancy.

Metamorphic 145
A type of rock that is formed from other rocks which have been heated and compressed.

Microphone 181
A piece of electrical equipment which detects sounds and converts them into electrical signals.

Mixture 102
An impure substance made from more than one element or compound, which can usually be separated by physical processes like filtration and evaporation.

Molecule 90, 102
A cluster of atoms chemically joined by covalent bonds.

Moment 171
The turning effect of a force.

Motor 191, 198
A machine that converts electrical energy into mechanical energy in a circuit.

Muscle 28
Tissues found in most animals that can contract and relax to move the animal or its internal organs.

Muscular system 28
The muscles that contract and relax to allow you to move your body or internal organs.

Natural selection 85
The gradual process in which characteristics become more or less common, which leads to evolution.

Nectar 56
Sugary solution formed by plants to attract insects to their flowers for pollination.

Neutral (pH) 121
When a solution is neither acidic nor alkaline, and therefore has a pH of 7.

Neutral (charge) 121
When a particle or object has neither a positive or negative charge.

Neutralisation 122
When an acid reacts with another substance (often an alkali) so that it is no longer acidic.

Neutron 95
A neutral particle that is found in the nucleus of atoms.

Noble gases 132
The elements in group 0 of the periodic table: helium, neon, argon, krypton, xenon, radon.

Non-contact forces 172
Forces that act between two objects that are not touching, e.g. Gravity.

Non-luminous 202
Objects that do not create light.

Nucleus (chemistry) 95
The central part of an atom, which contains the protons and neutrons.

Obesity 40
A word used to describe someone who is very overweight with a large amount of body fat.

Opaque 183
A term used to describe materials through which no light waves can pass, so you cannot see through them.

Optic nerve 185
The part of your nervous system which transfers electrical signals converted from light waves in your eye to your brain.

Organ 21
Groups of different tissues in the same place that complete the same function.

Organ system *21*
Groups of organs that work together to complete the same function.

Ossicle bones *180*
Three tiny bones that transfer vibrations from your ear drum on to your cochlea.

Ovary (animal) *48*
An organ in the female reproductive system that releases an ovum every 28 days.

Ovary (plant) *55*
Part of the carpel (female part of a flower) which contains plant ova.

Ovum *50*
(plural ova) A female reproductive cell.

Oxidation *116*
When a chemical reacts with oxygen or gains oxygen atoms in a chemical reaction.

Penis *49*
The part of the male reproductive system that passes urine from the bladder and becomes erect to ejaculate millions of sperm into the vagina during sexual intercourse.

Period (biology) *50*
Another term for menstruation, the monthly shedding of the lining of the uterus.

Period (chemistry) *132*
A horizontal row of elements in the periodic table.

Periodic table *131*
A list of all known elements in order of their atomic numbers, which is organised so that elements with similar chemical properties are arranged in vertical groups.

Peristalsis *31*
The rhythmical contraction of the muscles that line your digestive system to push food along it.

Permanent magnet *196*
A magnet that has its domains aligned and so will remain magnetic.

Pesticide *73*
A chemical used to kill pests.

Phloem *64*
Plant tissue that transports glucose made in photosynthesis from the leaves to the rest of the plant.

Photosynthesis *61, 113*
The process completed by plants and algae which uses light energy from the Sun to convert carbon dioxide and water into glucose and oxygen, and stores this energy in glucose.

Physical change *97*
A change that can be easily reversed and does not involve the formation of new chemical substances.

Pinna *180*
Your ear flap, which collects sound vibrations from the surrounding air to help you to hear.

Placenta *52*
The organ that develops in a pregnant woman to connect the umbilical cord of the baby to the wall of the uterus and carries food and oxygen to the baby and removes waste products.

Pole *196*
The end of a magnet where the magnetic field is strongest.

Pollen *55*
The male sex cell (gamete) of plants.

Pollination *56*
Reproduction in plants, which occurs when the male pollen meets the female ovum in a flower.

Polymer *143*
A very long chain molecule that is made up from many smaller units.

Potential difference *193*
A measure of how much electrical energy is carried by the current in a circuit. Potential difference is measured in volts, V.

Power *157*
The rate of change of energy transferred in a process. Power is measured in watts, W.

Precise *6*
How similar different measurements of the same thing are to each other.

Prediction *12*
A statement about the outcome of an experiment which is made before the investigation.

Pressure *92*
The force applied per unit area. Pressure is measured in pascals, Pa, or newtons per square metre, N/m^2.

Producer *71*
A term given to plants that are at the bottom of almost all food chains that obtain the energy from photosynthesis which is then passed along the chain.

Products *110*
The chemicals that are made in a reaction.

Property *94, 128*
A word that can be used to describe a substance; for example, 'good conductor of heat'.